DATE DUE

			PRINTED IN U.S.A.

COLLINS

Guide to the
BOTANICAL GARDENS

3/15/89

OF BRITAIN

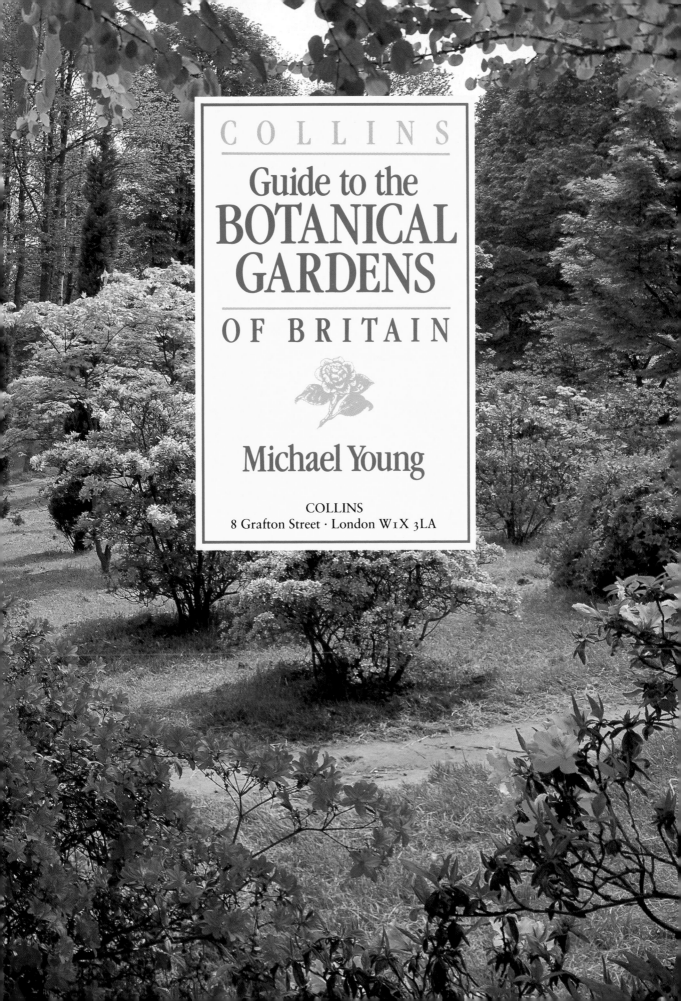

COLLINS

Guide to the
BOTANICAL
GARDENS
OF BRITAIN

Michael Young

COLLINS
8 Grafton Street · London W1X 3LA

First published 1987 by
William Collins Sons & Co Ltd
London · Glasgow · Sydney
Auckland · Toronto · Johannesburg

© Michael Young 1987

British Library Cataloguing in Publication Data
Young, Michael
Collins guide to the botanical gardens of
Britain
1. Botanical gardens – Great Britain –
Guide-books 2. Great Britain – Description
and travel – 1971 – – Guide-books
I. Title
580′.74′441   QK73.G7
ISBN 0 00 218213 0

Edited by Susan Conder
Designed by Janet James
Picture research by Sheila Corr

Photoset by Rowland Phototypesetting Ltd
Bury St Edmunds, Suffolk

Printed and bound in
Italy by New Interlitho SpA, Milan

Half-title page: *Cruickshank Botanical Garden.*

Title page: *Winkworth Arboretum.*

Contents

Foreword/6
Introduction/7
Map/15

Bath Botanical Gardens/17
Bedgebury National Pinetum/20
The Birmingham Botanical Gardens/25
Bristol University Botanic Gardens/31
Cambridge University Botanic Garden/36
The Chelsea Physic Garden/45
Cruickshank Botanic Garden/52
Dundee University Botanic Garden/55
Durham University Botanical Garden/58
Fletcher Moss Botanical Gardens/62
Glasgow Botanic Gardens/65
Harlow Car Gardens/71
Hull University Botanic Garden/76
Hume's South London Botanical Institute Botanic Garden/79
Leicester University Botanic Garden/83
Liverpool University Botanic Garden/88
Logan Botanic Garden/93
Oxford University Botanic Garden/95
The Royal Botanic Garden, Edinburgh/103
The Royal Botanic Gardens, Kew/113
Sheffield Botanical Garden/121
Southampton University Botanic Garden/128
St Andrews University Botanic Garden/131
University College of Swansea Botanic Garden/135
Wakehurst Place/137
Westonbirt Arboretum/143
Winkworth Arboretum/147
Younger Botanic Garden/151

Bibliography/157
List of gardens generally not open to the public/158
Acknowledgements/158
Index/159

Foreword

The first question any book purporting to be about botanic gardens must ask is what is the definition of such a garden? The answer is straightforward enough: a botanic garden is one in which the main concerns are instruction and research and where the plants are gathered together to form a scientific collection. In this very important respect a botanic garden goes beyond those in which the primary consideration is one of aesthetics or the creation of a pleasant environment.

Inevitably there will be some botanic gardens omitted from these pages, either because they do not conform to this definition or because they are gardens no longer worthy of being called botanic. Some are excluded because they have ceased to be scientific collections as such and have, over the years, been incorporated into local authority parks. Both Belfast and Liverpool City gardens are examples of those which have lost the scientific impetus required for inclusion in this book. Still other botanic gardens, such as those developed in recent years as seaside attractions, are omitted because their primary purpose is one of leisure.

Ease of public access is a further requirement for inclusion. At only one garden listed in the main text is an appointment to visit necessary and this itself is a recent introduction. Where access is generally difficult or non-existent there seemed little point in including the garden in the main body of the text. Nine university gardens fall into this category and are therefore listed separately at the back of the book (see p. 158).

Some other botanic garden may have been omitted through ignorance rather than design and for no other reason than that it is not known to me. However, nearly all those gardens listed have been visited by me during the eighteen months I spent preparing this book and I would like to thank the many garden curators, directors and their staff who gave generously of their time on my often unannounced arrival.

To my colleague at *The Times* newspaper, John Carey, who read the manuscript and without whose many suggestions and corrections this book would be the poorer, I am indebted. I should also like to thank Sheila Corr, whose photographic research has added greatly to the final appearance of the book, and Caroline White, my editor at Collins. Finally I would like to thank my wife for typing the manuscript and for adopting the role of a single parent while the work was in progress.

Michael Young. London 1987

Introduction

Many gardens are much too important and interesting to be left to gardeners. Deep in their soil lies a rich vein of history waiting to be tapped and enjoyed by people who would never themselves dream of picking up a trowel. On the surface of these gardens is a beauty that is capable of bringing peace to the most troubled and restless soul.

Nowhere is this more true than in Britain, where there are gardens that literally take the breath away. At the same time, these gardens, especially the botanic gardens and the tales that go with them, provide a fascinating glimpse into the social and political development of Britain over many hundreds of years. The history of botanic gardens includes travellers and adventurers, who scoured the globe in search of new plants; kings and commoners, priests and publicans, academics and artisans, who were brought together in a joint endeavour to create something to treasure, sometimes in the most unpromising or bizarre circumstances. In certain cases, they worked in the wilderness, surrounded by scenery of untrammelled wildness; elsewhere their task was to soften the harsh edges of the Industrial Revolution by creating oases of green space in the grimy cities.

It is not just the past that comes to life in these places. The battles that these pioneers fought are mirrored in the crises facing many major gardens today: crises of cash and conservation. Botanic gardens, earlier known as physic gardens, have existed for many centuries. Their original purpose was to grow herbs for medicinal application. Aristotle kept a physic garden; so did Pliny. Dioscorides, author of *De Materia Medica*, the herbal that influenced European thinking from the first to the sixteenth centuries, must also have possessed one. His much copied and translated work is a curious blend of sharp observation and folklore.

Since the sixth century most monasteries in Britain had a herb garden for both culinary and medicinal purposes. Indeed monasteries were often at the centre of the dissemination of medical knowledge, and their herbals, although largely based on Greek originals, do display a first-hand familiarity with many of the plants they describe. One such is the *Glastonbury Leechdome*, written in the tenth century, in which the plants' 'vertures' cover both animals and people alike.

Exploration during the fifteenth century had a profound effect on the advancement of medical knowledge. The Portuguese advanced along the coast of the African continent, eventually rounding the Cape in 1488. Columbus boldly struck out into the Atlantic in 1492 and as the century closed Amerigo Vespucci charted the coast of South America. One of the benefits of such exploration was that an increasing number of plants were brought back to Europe. They found their way into the botanic gardens that were established in Europe during the sixteenth century. The botanic garden of Padua was the earliest, founded in 1545 close to Venice, where the university already enjoyed a considerable reputation in natural history. Other European cities followed; Leyden in 1587, Montpellier in 1596 and Paris in 1597.

The Dean of Westminster, William Harrison, writing in 1577 in *A Description of England*, describes the Elizabethan enthusiasm for plants from the known world, 'from the Indies, Americas, Canary Isles'.[1]

Two particularly significant years for English botany were 1596 and 1597. John Gerard, the barber-surgeon and gardener to Lord Burleigh, issued a list of the 1000 or so plants growing in his own garden at Holborn in London, near what is now Fetter Lane. This catalogue records the first mention of many plants introduced into the British Isles. In 1597 Gerard published his best-selling and now famous *Herball*. Although exhibiting a first-hand knowledge of plants the book was in fact a bit of a cheat. It was a translation of a herbal published in 1583 by the well known Belgian botanist, Rembert Dodoens. Any best-selling book of the day was aimed at gentlewomen and Mrs Gerard's contribution was in advising her husband about what exactly women like to read. For example, listed under 'The Virtues for Marigolds' was both medicinal advice and the following: 'the floures are dried and . . . put into broths . . . in such quantity, that in some Grocers' or Spice-sellers' houses are to be found barrels filled with them, and retailed by the penny more or lesse, insomuch that no broths are well made without dried Marigolds'. With the addition of a job lot of wood-block illustrations picked up cheap on the Continent, the book was passed off as being by Gerard's own hand.

Never slow in coming forward, Gerard had nine years earlier penned a letter to the Chancellor of Cambridge University, suggesting himself as the ideal person to tend a botanic garden. Whether the letter was even sent, we do not know; certainly Cambridge did not get their garden for another 172 years and the distinction of making the first botanic garden in the British Isles went instead to Oxford.

The Earl of Danby's purpose in founding the physic garden in 1621 in Oxford on ground leased from Magdalen College was clear enough. He was 'determined to begin and finish a place whereby learning, especially the faculty of medicine, might be improved', at a time when medicine and botany were inseparable, and where, 'divers simples might be devised for the advancement of the faculty of medicine'. (Simples are, according to the *Shorter Oxford English Dictionary*, 'a medicine or medicament composed or concocted of only one constituent, especially of one herb or plant'.)

This simple beneficial principle also lay behind the founding of the physic garden at Edinburgh in 1670. Here Robert Sibbald, the first Professor of Medicine at Edinburgh

John Gerard, whose 'Herball' was published in 1597.

University, and his friend, Dr Andrew Balfour, sick to death of quack medical practitioners and their fraudulent concoctions, began to grow medicinal herbs on a small plot of land.

Three years later the Society of Apothecaries in London founded their physic garden at Chelsea as a living textbook where students could learn taxonomy and thereby the medicinal properties of the plants.

Naturally enough, the botanic gardens became the homes of plants that were either exotic or difficult to grow. Both kinds attracted the attention of writers such as Thomas Johnson and John Evelyn. Johnson has a delightful description of a bunch of bananas hung in his shop in Snow Hill (see page 51). Evelyn, always a prolific writer, describes, in a letter dated August 21, 1668 to the Earl of Sandwich, how 'the Planta Alois, which is a monstrous kind of Sedum, will like it endure no wett in Winter, but certainly rotts if but a drop or two fall on it, whereas in Summer you cannot give it drink enough'.

Evelyn was also something of an expert gardener, constantly visiting the botanic gardens around Europe. He was often at Oxford and in 1644 visited Padua, 'rarely furnish'd, with plants, and gave order to the gardener to make me a collection of them'. In the same letter to the Earl of Sandwich he also writes of his attempts to grow 'Pome-granad'. 'They will flower plentifully,' he says, 'but beare no fruit with us.'

During the seventeenth century the sciences of medicine and botany began to diverge. Thomas Johnson's corrected and expanded edition of Gerard's *Herball*, published in 1636, demonstrates not only an acute observation of plants but also a real love in their

The title-page from John Parkinson's 'Theatrum Botanicum', published in 1640.

description. John Parkinson, apothecary to James I, published *Paradisi in sole Paradisus terrestris* in 1629 and *Theatrum Botanicum* in 1640. These books, although strictly speaking herbals, have a gardener's passion for plants ringing through every page. The books fired the imagination of John Ray, who entered Cambridge University in 1644 and went on to embrace British natural history in his work published in 1686, *Historia Plantarum*. In it he mentions a number of relatively new introductions growing in what must have been his own garden in Cambridge.

The speed at which the study of botany as a science developed can be gauged by the fact that in 1724 Philip Miller, Gardener at the physic garden at Chelsea, published his two-volume *The Gardeners' and Florists' Dictionary*, followed seven years later by his famed *The Gardeners' Dictionary*. The latter was inspired by Miller's determination to disseminate the knowledge he had gained at Chelsea in cultivating the plants flooding in from places such as the West Indies, North America and even Siberia.

Not until 1760 did Cambridge get its botanic garden, the year after Augusta, Dowager

Princess of Wales and mother of George III, began her nine-acre (3.6 hectare) botanic garden at Kew. The University of Cambridge botanic garden was founded by Richard Walker, Vice Master of Trinity College, who saw it as a traditional physic garden where 'experiments shall be regularly made and repeated, in order to discover their (plants') virtues, for the benefit of mankind'.

This view prevailed at Cambridge until 1825 when the garden was taken over by the young botanist, John Stevens Henslow. In his much quoted address to the members of the University of Cambridge on the *Expediency of Improving, and on the Funds Required for Remodelling and Supporting the Botanic Garden*, published in 1846, Henslow defines the modern purpose of a botanic garden. Large numbers of plants should be grown not 'merely for systematic improvement, but for anatomical and other experimental researches essential to the progress of general physiology'.

Henslow also stressed the importance of protecting in the long term the live data base which botanic gardens inevitably become. 'It is impossible to predict what . . . species may be . . . dispensed with . . . without risking some loss of opportunity which that species might have offered to a competent investigator.'

All this time the Royal Botanic Gardens at Kew under Sir Joseph Banks had been continuing its exploratory work establishing an international reputation as *the* garden to which collectors and botanists sent seeds and specimen plants.

The nineteenth century also witnessed the birth of gardening as a leisure activity on a grand scale, with writers such as John Claudius Loudon publishing many thousands of words popularizing the subject. In 1826 Loudon launched *The Gardeners' Magazine*, the first periodical aimed at gardeners from all social classes. The aim was to 'disseminate new and important information on all topics connected with horticulture, and to raise the intellect and the character of those engaged in this art'.[2]

Civic pride was also a commodity much in evidence. Any city worth its salt had to have its own botanic garden. Most were founded by private subscription raised among the wealthy industrialists who found they had time and money enough to indulge their hobby of gardening. Some floundered almost immediately; others, including those in Glasgow and Bath, survived to become public parks. A few, such as Liverpool's botanic garden, have lost their identity altogether – and one, established in Birmingham in 1829, remains financed by private subscriptions.

The Victorian love of novelty played a critical role in these botanic gardens, as did the development of conservatory constructions. The development of new techniques in the manufacture of glass and iron, and later the repeal of the glass tax in 1845, led to conservatories as we know them today: vast glass enclosures in which plants were protected during the winter months and complete micro-climates were created. Joseph Paxton was among the most ambitious of the early builders. In 1839 he created a monument to Victorian ingenuity with his vast conservatory at Chatsworth. Almost simultaneously the exquisite curvilinear palm house at Bicton in Devon was made.

Paxton's conservatory at Chatsworth decayed and was pulled down but Decimus Burton's elegant palm house, built in 1848 at Kew, remains a monument to Victorian extravagance.

The number of exotics that could be grown increased and huge conservatories were a prerequisite of any botanic garden. Some, such as the Kibble Palace erected at Glasgow in 1871, became attractions in their own right. Others, such as the huge circular extravagance designed by Loudon for the Birmingham Botanic Garden, proved prohibitively expensive to build and cheaper, and it must be said, more staid alternatives were built instead.

The Victorian botanic gardens demonstrated the wonders of the plant kingdom and the special skills needed to cultivate plants from around the globe. Their educational role was one of teaching by example through the development of good gardening techniques. The role of the traditional botanic garden was laid to rest while that of botanical excellence, diversity and novelty, gained the upper hand.

Developments at Kew and Edinburgh and, to a lesser extent, Chelsea continued. Both Kew and Edinburgh played their part in the introduction of new species to cultivation in Britain and both acted as staging post and distribution points for many plants. Kew's pre-eminence has been maintained and strengthened through the twentieth century while universities around the country have been establishing new botanic gardens. Many have been developed from earlier Victorian gardens while others have been created from open fields.

Albert Forbes Sieveking says in his introduction to Sir William Temple's *Gardens of Epicurus and other Seventeenth Century Essays*, published in 1908, 'There would seem to be some subtle psychological nexus between the Garden spirit and the Soul of Universities and Academies – the classic and sacred Groves of Thought, Learning and "Impassioned Contemplation" and perhaps it is well that when the Schoolmaster and Professor are abroad – the spirit and soul of Gardens should be also alive and active.'

It is fitting that the practice of establishing and maintaining botanic gardens, traditionally part of university life, should have been confirmed by the universities of the twentieth century. Here plant collections flourish and in the process of cultivation wonderful environments are created. Henslow at Cambridge wrote in 1846: 'The larger the number of species that are cultivated in a Botanic Garden, the greater will be the facilities afforded us all'. He had in mind scientific benefits, of course. But he could also have been referring to the value of botanic gardens as public amenities. For although their primary purpose has always been scientific, and this is how they are still seen by most curators, maturity has turned them into areas of repose and retreat that can be shared by scientist and layman alike.

The sad fact is, however, that they are under-used amenities. Not only have most of them become beautifully mature gardens but nearly all of them display and label their plants in the most clear and instructive way possible. Every effort should be made to entice the public through the gates, but all too often a singular lack of imagination is shown. One definition of a botanic garden which I have come across is 'a garden to which the public have some access and in which the plants are scientifically labelled'. This limited view prevails among the hierarchy of many of the gardens' directors.

This became clear when I asked a number of directors and curators how they saw the role of a botanic garden in today's world. Professor Arthur Bell, Director of the Royal Botanic Gardens at Kew, was unequivocal. He sees Kew as a reservoir of living plant material and a dry and live data base for research. Though he sees Kew's main role as scientific, this must, in his opinion, include the more traditional roles of identifying economic plants and pursuing taxonomy.

Kew's scientific data has had many unforeseen spin-offs. Its vast collection of microscopic slides covering every aspect of wood enables staff to identify wood from almost any source and in almost any form. In the past they have been called in to identify bits of tree roots suspected of causing house subsidence and have even identified minute traces of sawdust for Scotland Yard. Third World countries constantly use Kew as a middleman or go-between for timber-breeding, crop research and food development programmes.

Professor Bell trained as a biochemist and cites with some enthusiasm Kew's 137,000 index cards on economic plants as an untapped well of information. To tap that information would be to find new uses for old plants at a time when only about one per-cent of all plants have actually been studied for their economic properties.

Duncan Donald at the Chelsea Physic Garden would also like to see more projects that exploit the medicinal and economic properties of plants and regards Chelsea as being ideally placed to supply authenticated specimens for clinical trials.

He is actively maintaining links with scientific institutions; Chelsea College has in the recent past used feverfew in a clinical study of migraine. Donald feels Chelsea's future lies in developing a strong educational role, teaching visitors about plants and their properties through detailed labelling and by establishing a positive schools policy. From a purely pragmatic point of view, he has to attract visitors to make the garden self-financing.

No such commercial pressure is brought to bear at Oxford University Botanic Garden. Here the principal role, as at any university botanic garden, is to aid teaching and research within the university. At the beginning of each term the lecturers produce a shopping list of requirements that must be met by the garden staff. The whole of the plant collection is geared towards meeting this demand. The aesthetics of the garden are of course considered but in a secondary role.

Cash problems are not new for botanic gardens. John Stevens Henslow, in his 1825 address to the members of Cambridge University, said: 'Unless an adequate sum can be secured . . . we shall run the risk of following the example of some provincial establishments, which have flourished for a while under the excitement of novelty, but which have ultimately fallen into neglect, or even been sold to pay off debts that could not be met under the diminished amounts of annual subscriptions'.

Today some botanic gardens find themselves facing a real crisis. The University of St Andrews Botanic Garden, for example, is under threat, and Bob Mitchell, its curator, voices sentiments which echo Henslow's. In the face of threatened economies that will force the sale for housing development of part of the garden, Bob Mitchell articulates the over-riding value and relevance of botanic gardens in a world criss-crossed by instant communication and technology. A botanic garden should be regarded as a growing gene pool of material for academics to use at any time in the future.

Another area of profound concern among the directors of botanic gardens is one of conservation. Many notable plants threatened with extinction have been saved by the intervention of botanic gardens. Although botanic gardens are ideally suited for growing and preserving rare specimens, a rare plant once brought into a botanic garden is at risk and remains just one plant from a colony, just one sample of the gene pool. Ken Burras at Oxford endorses this view and believes a more practical approach to conservation is in alerting the public to the preservation of a plant's natural habitat.

It is a view shared by Professor Bell at Kew, where they hold many rare plants and are always concerned to encourage other botanic gardens throughout the world to advise governments on the preservation of natural habitats. But Bell believes that to rescue the plant and then to destroy the habitat is an admission of failure.

The scientists at today's botanic gardens are helping to turn deserts green, are finding commercial uses for plants hitherto unthought of and are finding a new role in the conservation battle. As a by-product they have created, or are the custodians of, beautiful environments from which we as laymen can only benefit. The plants they nurture are grown in near perfect conditions and present a paradise for the plantsman. Such environments provide areas of seclusion and repose in a world where space and greenery are increasingly threatened. We as the visiting public should cherish these gardens and should encourage their development and their protection, for, 'Everywhere in Life and Letters the Spirit of the Time and of the Garden walk hand in hand and cast their spell over our souls and bodies'.[3]

1. Miles Hadfield, *A History of British Gardening*.
2. Gerard, *The Herball*, 1636 edition.
3. R. T. Gunther, *Oxford Gardens*, Oxford, Parker and Son, 1922.
4. Sir William Temple , *The Gardens of Epicurus, with other XVIIth Century Garden Essays*, Chatto and Windus, London, 1908.
5. S. M. Walters, pp. 41–2, 'Statutes and Orders', *The Shaping of Cambridge Botany*, Cambridge University Press, 1981.
6. From the preface to the first completed volume of *The Gardeners' Magazine*.
7. Ibid.

Cruickshank, Aberdeen ■

Dundee ■
St Andrews ■
■Younger

■Glasgow   ■Edinburgh

■Logan

Durham ■

Harlow Car, Harrogate ■

Hull ■
Ness ■   ■Fletcher Moss, Manchester
Sheffield ■

■Leicester
■Birmingham

Cambridge ■

■Swansea

Oxford ■
■Westonbirt
■Bristol
■Bath   Chelsea Physic Garden ■
Kew ■   ■South London Botanical
                Institute
Winkworth ■
Southampton ■   Wakehurst Place ■   ■Bedgebury

Early postcard views of Bath Botanical Gardens.

Bath Botanical Gardens

Royal Victoria Park · Marlborough Lane · Bath · Avon · BA2 2NQ
Tel. 0225-24728

There is a great deal of civic pride vested in Bath's botanic garden, now approaching its centenary year. Although it occupies a mere seven acres (2.8 hectares) within the Royal Victoria Park, it claims to have one of the finest collections of plants growing on limestone in the south of England. The garden was laid out in 1887 to receive a gift of 2,000 plants from a Mr C. E. Broome of Elmhurst, Batheaston. The Corporation, with the aid of subscribers, raised enough money to prepare the ground on the site of an earlier 1840 botanic garden which had foundered due to lack of support.

The garden was laid out to form an attractive and picturesque feature within the Royal Victoria Park rather than as a garden with rigid and scholarly botanical leanings. In 1926 a small stone pavilion from the British Empire Exhibition held at Wembley two years earlier was moved to the site and erected on a terrace overlooking the water feature.

Today the garden has the feel of a well tended public park. However, it is full of interesting plants and the limestone rock garden offers a stunning display of colour from early spring. The garden also supplies botanical specimens to the University of Bath and exchanges seed lists with botanic gardens around the world.

The garden's policy today is one of public education by example. A fine collection of plants is maintained in good condition for the perusal of serious gardeners as well as the interest of the general public. The existing water feature is currently being developed; a stream now runs down from the central pond and planting on the banks is becoming dense and lush. The garden also boasts a magnificent herbaceous border, the history of which, like that of the rest of the garden, is obscure. Unfortunately no plans of the garden's early layout have survived.

Serpentine paths weave their way across the lawns where many specimen trees have grown to ripe maturity. There are a number of island beds and mixed shrub borders where liberal use is made of ground cover plants such as geranium and epimedium. So dense is the planting in parts of the garden that often a sub-tropical feel is evoked.

There are a large number of variegated plants such as *Cornus* mas 'Variegata' growing in happy familiarity with a wide and spreading *Prunus* 'Shirotae' and *Acer palmatum*.

Indeed there are many maples, the most attractive of which is the low rounded bush *A. palmatum* 'Dissectum Ornatum'.

Close by the western entrance to the garden is a striking *Yucca gloriosa*, its limbs contorted and snake like. The lawn here is dominated by an impressive *Catalpa bignonioides* 'Aurea', its leaves golden in the sun and its long limbs now propped for support. There is, too, a beautiful black mulberry, *Morus nigra*, its fruit like fat, succulent raspberries, and a tall stately sweet gum, *Liquidambar styraciflua*.

The limestone rock garden is a delight in spring, when it comes alive with flowering bulbs and alpines. Later a mass of geraniums join forces with *Astrantia major*, whose flowers look like tiny Elizabethan ruffs. But even here it is difficult to ignore the mature trees that dominate the garden. A *Metasequoia glyptostroboides*, known only from fossil remains until in 1941 a living specimen was found in China and the seeds sent back for distribution, soars skyward alongside a *Sequoiadendron giganteum*.

By the water more maples produce a kaleidoscope of colour. Hostas, agapanthus, *Rodgersia podophylla* and *Primula florindae* all grow dense and rich while later in the year, ligularia sends up its vivid orange-yellow flowers. *Dipelta floribunda* can be found here too, a fragrant flowering shrub that repays occasional hard pruning and seems to shed its bark like blistered skin.

The vivid display of summer bedding is confined to a small round bed. Far more attractive is the long, deep herbaceous border, where all the old garden favourites are to be found: *Erigeron*, with its glaucous leaves topped by poppy-like flowers; a purple-stemmed filipendula; echinops; monkshood; the lovely white-flowering *Achillea ptarmica*; ox-eye chamomile, *Anthemis tinctoria*; and that most beautiful of British natives, *Lythrum salicaria*, the purple loosestrife.

The garden also possesses some very fine magnolias, including *M. stellata*, *M. dawsoniana* and, most elegant of all, *M. tripetala*, whose pendulous branches and umbrella-like clusters of leaves give the magnolia its common name, the umbrella tree.

The terrace and water feature at Bath Botanical Gardens.

Bedgebury National Pinetum

on the B2079 between Goudhurst
and Flimwell · Kent
Tel. 0480-211602

Open every day,
10 am–8 pm or dusk.
Entrance fee.

John Evelyn, the great seventeenth-century arboriculturist and author of *Sylva*, the classic work on trees published in 1662, would have loved Bedgebury Pinetum in Kent. The hundred-acre (forty-hectare) site is one of exquisite natural beauty with two stream-lined valleys, separated by a ridge converging on a large lake. It is the slopes of these valleys, thick in irregular plantations of conifers, that would have fired Evelyn's imagination.

The pinetum occupies a small fraction of land on the west edge of Bedgebury Forest, which has existed in one form or another for many centuries. In Tudor and Stuart times it supported both the iron-smelting and the charcoal industries. More recently it supplied much needed timber for Second World War ships.

When the Forestry Commission and the Royal Botanic Gardens at Kew acquired the site for a pinetum in 1924, it consisted mainly of Scots pine, coppiced chestnut and oak, with just a scattering of exotic conifers. Since 1945 the Forestry Commission have had sole responsibility for what has become the most comprehensive collection of conifers in Europe, with more than 200 species represented.

The aims of the pinetum are clear: to collect together cone-bearing trees from most of the temperate regions of the world and make them available for scientific research; to grow specimens for visitors to evaluate their decorative worth; and to test newly introduced species.

Bedgebury has a wild beauty which defies any attempts to impose a convenient mental order on the site. Broad turf rides hug the contours of the undulating site and cross the pinetum like the bones of a fish. One can strike off from these main paths whenever the fancy takes hold. Broad-leaved trees add brilliant accents to the solid acres of evergreen, the variety of which can at times seem overwhelming.

One's first impression as the land falls away from the car park is of a solid green phalanx of every conceivable shade upon which is splashed the blue hues of cedars. In autumn the colours of the broad-leaved trees and deciduous conifers add a dramatic intensity. As the path drops down, an avenue of purple and deep red liquidambar is a bright feature on the landscape. A group of *Pinus sylvestris*, tall stately pines with their tawny bronze limbs, stands out among the evergreen conifers while the occasional beech, alive with golden leaves, adds a dab of Impressionist colour. Everywhere is the most incredible smell of pine and underfoot is the springiness of compacted needles. A larch plantation creates a golden haze, with deciduous needles falling in the breeze, and great thickets of rhododendrons cross the open terrain.

From the top of Hill's Avenue the view is long and straight: a wide sweep of soft, springy turf edged by many different conifers. Some are tall and fastigiate, others squat or carrying their branches down to the ground, like the curtains beneath a proscenium arch.

Bedgebury Pinetum has become the most comprehensive collection of conifers in Europe.

The Japanese black pine, a tall, masterful tree, is close by while at the end of Hill's Avenue the view is sealed by a copse of spruce. To the north runs North Avenue where *Cephalotaxus, Saxegothaea conspicua* and the nutmeg, rare genera related to the yew, can be found. A clump of tall red fir, *Abies magnifica*, with grey bark and glaucous needles, stands alongside an avenue of Leyland cypresses, the foliage of which is so dense that their trunks are virtually obscured. Hill's Avenue plunges down to a stream before rising gently and running off to the copse of spruce. Here by a small waterfall is a clump of small hump-back *Chamaecyparis lawsoniana* Hazelmere Form.

Cypress Valley follows the course of the stream cutting across Hill's Avenue. Just beyond the waterfall are two rare *Chamaecyparis thyoides*, their tiny cones dabbed with blue. From here up a steep slope climbs an escarpment of low growing yew trees whose colours in autumn range from deep green to vivid yellow.

Also on this slope is a fenced area dominated by a tall oak. Within the fencing is displayed a range of slow-growing conifers including many forms of *Thuja, Picea* and *Juniperus*. They grow little more than an inch or two (2.5 to 5 centimetres) each year, the new growth appearing like fresh plumage at the end of the branches.

Below the yew bank, a walk crowded with rhododendron follows the course of the stream to Marshall's Lake. Along this walk grows the giant redwood *Sequoiadendron giganteum*. A native of the western slopes of the Sierra Nevada, California, it can live for 3,000 years and reach a height of 300 feet (90 metres). Giant redwoods were introduced into Britain in 1853 and have soft spongy bark.

In autumn, the water of Marshall's Lake can be dark and uninviting under a dour sky. Even so, an avenue of swamp cypress, *Taxodium distichum*, creates rich brown reflections in the water, while their roots enjoy the moisture at the edge of the irregularly shaped lake. Nearby a *Rhododendron auriculatum* has grown tall, discarding in the process the density one usually associates with this genus.

Above the lake on a rise of land which soon becomes relatively flat, are the forest plots begun in 1929 to complement the pinetum. There in grid-like fashion, conifers and deciduous trees each occupy quarter-acre (.10-hectare) plots over a fifty-acre (twenty-hectare) spread. The wide avenues offer sweeping vistas and painterly chiaroscuro perspectives. This is a place to encounter the true silence of the forest, shattered only by the occasional guttural call of a pheasant.

Few people penetrate this quiet, dense place where the common oak, *Quercus robur*, clings to its leaves well into autumn and the foliage of the red American oak, *Quercus rubra*, adopts the colours of fire. There is a plot of *Metasequoia glyptostroboides*, the dawn redwood, thought extinct until discovered by scientists in China in 1941 and eventually introduced to Britain in 1948.

In the forest plots, the deciduous trees are shot through with light, and bracken around them flourishes on the forest floor. The conifers, by contrast, are densely planted, and the thick canopy of Oriental spruce, *Picea orientalis*, shuts out virtually all light.

Above Marshall's Lake, Bedgebury Pinetum.

Above *Contemporary view of terrace glasshouse and exhibition hall, which was erected in 1884.*

Below *Modern view.*

The Birmingham
Botanical Gardens

Westbourne Road · Edgbaston · Birmingham · B1J 3TR
Tel. 021-454 1860

Open every day except
Christmas Day:
9 am–8 pm Mon–Sat,
10 am–8 pm Sun (summer);
9 am–dusk Mon–Sat,
10 am–dusk (winter).
No entrance fee.

The heart of the Birmingham Botanical Gardens is often described as being the large expanse of south-facing lawn that drops down from the glasshouse terrace into a natural amphitheatre and on to the peripheral planting of borders and woodland. But the terrace itself is the real heart, a place to sit in the warm sun, to absorb the views and the kaleidoscope of life unfolding on all sides. The landscape dips down, across bog and water gardens and on to the distant trees of Chad Valley, creating the impression of a continuous landscape uninterrupted by any building. The view defies logic. It is difficult to believe that the centre of Birmingham is just over a mile (1.6 kilometres) away.

These botanical gardens have adapted to the twentieth century and have solved the identity crisis posed since the Second World War, namely how to fulfil the more traditional scientific needs of its Members while simultaneously retaining popularity as a public amenity. Today it remains the only provincial botanic garden in the country still owned and managed by a private Society. The garden's educational facilities have been extended to include children and the Society has been turned into an Educational Trust. This ability to adapt is one that has saved the gardens more than once in their history.

The idea of a botanic garden in Birmingham was first mooted in 1801 when a meeting convened to discuss the foundation of a Birmingham Botanical and Horticultural Society. This early attempt to raise subscriptions foundered and the Society did not finally get under way until 1829. The sixteen-acre (6.4-hectare) gardens were not acquired and formally opened until 1832 on land formerly occupied by a farm. At the time it had become almost *de rigueur* for a large provincial city to acquire a botanic garden through subscription to promote and encourage the very natural interest in exotic flora engendered by the expansion of the British Empire. Already, Liverpool, Manchester, Hull and Glasgow possessed such gardens. The professed aim of the Birmingham Society was to give its subscribers access to the greatest possible range of plants that were currently flooding into the British Isles, which would flourish in the open or under glass.

John Claudius Loudon, the great propagandist for the study of gardening among the middle classes at the time, was invited to lay out the gardens in a way that combined a 'scientific with an ornamental garden'.[1] Loudon's design was accepted with reservations. The spectacular circular glasshouse Loudon considered so essential to his overall scheme was deemed by the Society far too expensive. The subsequent change to a cheaper and more conventional glasshouse range designed by someone else and grafted on to the original scheme incensed Loudon, who included his plan in later editions of his best-selling *Encyclopaedia of Gardening* as an example of a well designed small botanic garden.

In June 1832 the gardens were opened to shareholders and their families for an annual subscription of one guinea. Gifts poured in from around the world. These included more than 1,370 packets of seeds, a dozen boxes of orchids from Brazil supplemented by gifts from Kew, Edinburgh and Chelsea, and donations of plants coming from as far afield as the Calcutta Botanic Garden. Such largesse is a characteristic of gardeners and today Birmingham, like many others, still exchanges seed distribution lists with international botanic gardens.

By 1834 the gardens had more than 9,000 species and were soon engaged on an active seed distribution programme to subscribers. Although plants in the gardens flourished its finances were far from healthy. Having begun its life with a deficit, a further share issue was thought necessary in 1836 and later, in 1840, the annual subscription was raised to £1 11s 6d.

The parlous financial state of the gardens necessitated a change in direction and over the next twenty years greater emphasis was placed on developing ornamental features in the hope of attracting more subscribers. One such development was to enhance the gardens' attractions through more decorative planting. In addition the fountain at the intersection of paths below the lawn was constructed in 1850; two years later the lily house was built to accommodate the exotic rarity, *Victoria amazonica*, given to the gardens by Chatsworth's head gardener, Joseph Paxton. The planting of popular rhododendrons and azaleas was also extended.

Nowadays from the terrace the gardens are seen to possess a quiet Victorian grandeur, with the main view focused on the 1873 bandstand which sits in the bowl of the natural amphitheatre. Around the lawn serpentine paths follow the slope's contours, taking the visitor gradually down the decline and into the mature planting which occupies the lower ground beyond the bandstand. All the paths eventually intersect at a fountain tucked in a gloomy glade out of immediate view on the edge of the woodland.

Over the fountain tower a huge *Cedrus deodara* and a tall *Magnolia acuminata*, the latter as sedate and comfortable as any native British plant. Below this junction, on a short but steep escarpment, is the rock garden. Built around a small spring in 1895, the rock garden contains 250 tonnes (tons) of Millstone grit from Yorkshire. In spring azaleas and magnolias flower above hundreds of bulbs and alpines. On the surface of the

The fountain at Birmingham Botanical Gardens, built in 1850.

pool float water lilies and the fascinating but diminutive water hawthorn, *Aponogeton distachyus*, its waxy white flowers reaching up above the leaves, which form tiny floating rafts. In autumn maples adopt rich tones of umber and gold. That most alluring of Himalayan plants, *Leycesteria formosa*, with its vigorous bamboo-like growth, hangs its claret-coloured racemes in such a way that its colloquial name, gypsy's ear-rings, becomes immediately obvious. Above the water dragonflies bob and weave at high speed.

A dense and thickly planted rhododendron walk runs from the rock garden along the southern perimeter of the gardens and terminates in a bog where spectacular clumps of *Gunnera* grow. Punctuating this walk are a host of interesting trees such as *Cercidiphyllum japonicum*, a coppice-like mass of branches sporting beautiful heart-shaped leaves

which turn bright pink in autumn. There is also a large *Acer griseum*, its bark peeling in long papery strips, and *Acer grosseri*, with a vivid green and silver striated bark.

Beneath the lush rhododendrons grow carpets of *Helleborus corsicus* and the native caper spurge, *Euphorbia lathyrus*. Considered little more than a weed by many, the latter is a delightful addition to any garden, with its ripe seeds bursting from the pods in high summer with an audible snap.

Above the bog the azalea walk ascends the slope among many herbaceous perennials. The specimen trees always make their presence felt. These include *Betula papyrifera*, the paper birch from North America. (Indians used its white waterproof bark flushed with patches of pink to make canoes.) Close by is an exquisite *Arbutus menziesii*, whose mature bark peels away, exposing bone-smooth pink surfaces, while the current season's growth is a rich, fresh sage green. Its subtle colouring makes the scarlet berries of the neighbouring mountain ash *Sorbus aucuparia* seem coarse.

There is a convenient shelter there against the rain. Nearby are raised beds of grey-leaved plants and through the woodland to the west is a children's play area, an aviary, a pool for waterfowl and a deep herbaceous border, backed by a low terrace. Sadly the border lacks conviction, but if the overall structure is unimpressive, the individual plants such as phlox, geranium, campanula and echinops are worth seeing. The saving grace, however, is the magnificent background supplied by an assertive copse of *Pinus nigra*. Their dark feathery foliage is carried high up in the tree and their relatively straight branches are outlined against the sky.

Sited discreetly in a corner to the south-west of the glasshouse terrace is the rose garden, a place rich in evocative smells which conjure up memories of China tea on warm summer evenings. The rose garden was replanned in 1904 when it assumed its formal rectilinear shape with central terrace lawns. Floribundas, hybrid teas and climbers can all be found there.

The glasshouses are the crowning glory of the Birmingham Botanical Gardens, and are crowded with a mutinous density of informative planting. The palm house contains plants of economic value: tea, coffee, citrus and cinnamon. The tall variegated reed, *Arundo donax*, reaches high into the roof space and mingles with the sky-blue flowers of perennial morning glory, *Ipomoea acuminata*. Huge palms reach the skylights and the saucer-shaped mauve flowers of *Tibouchina urvilleana* make clouds of colour. Epiphytic orchids grow on branches and Latham's tree fern, *Dicksonia* x *Lathamii*, is also housed here. This very rare plant is a hybrid, bred in 1872 when the gardens' curator, William Bradbury Latham, crossed *Dicksonia antarctica* with *Dicksonia arborescens*.

Central to the humid, steamy lily house is a pool where water lily leaves float on the surface while the flowers hoist themselves clear. At the pool side papyrus, *Cyperus papyrus*, stretches to the roof. *Dichorisandra thyrsiflora* dazzles with its deep-blue flowers carried on bamboo-like stems and vigorous sugar cane, *Saccharum officinarum*, bends over where it, too, touches the glass roof. The intensely beautiful palm-like

Inside one of the densely-planted glasshouses at Birmingham.

Ravenala madagascariensis produces enormous fan-shaped leaves while close by *Monstera deliciosa* has eye-catching soft yellow flowers, like those of the cuckoo pint, *Arum maculatum*. By far the most exotic plant is *Calliandra haematocaphala*, a small tree from southern Brazil. Its extravagant flowers are dazzling inflorescences of red hairs, each one tipped with black, like finite fibre optic cables. Its common name, Cupid's shaving brush, indicates how soft it is.

In the cool house the heating is just enough to keep winter frosts at bay. A large specimen of Peruvian thorn-apple, *Datura rosea*, seems happy with this arrangement, as does the yellow-flowered ginger lily, *Hedychium gardnerianum*, whose intense fragrance seems to infiltrate the pores of one's skin. *Datura sanguinea* grows through the top lights and into the fresh air.

In summer the terrace is a sheltered sunny spot. Here are the California tree poppy, *Romneya coulteri*; the Moroccan broom, *Cytisus battandieri*; and the glory bower, *Clerodendron trichotomum*, whose creamy jasmine-like flowers emerge from mauve calyces. Palms, cannas and succulents grow nearby in large cedar tubs.

1. pp. 407–28, *The Gardener's Magazine*, vol. viii, as quoted in Phillada Ballard's *An Oasis of Delight, The History of the Birmingham Botanical Gardens*, Duckworth, London, 1983.

Bristol University Botanic Gardens

Bracken Hill · North Road · Leigh Woods · Bristol
Tel. 0272-303030

Open Mon–Fri,
9 am–5 pm;
members only Sat, Sun;
closed Christmas week, Easter
week and Bank Holidays.
No entrance fee.

Bracken Hill House, which stands within the grounds of the University of Bristol Botanic Garden, was built in 1886 for the tobacco baron Melville Wills. It is a curious elongated building, a hotch-potch of styles: part in the style of Lutyens, part tile-hung, with a round baronial tower at one corner and elaborate plaster mouldings picked out in white on the gables. The house is now used for student accommodation. The paintwork is cracked and peeling and the windows exhibit a medley of indifferent curtains.

Both house and five-acre (two-hectare) garden were bequeathed to the University in 1959 by Captain Douglas Wills, son of Melville, on the understanding that the grounds were used for the study of agriculture. While the house is now run-down, the garden is certainly not: it is extremely well cared for with more than 4,000 different kinds of plants.

The site slopes gently to the north and is more or less dissected. The half surrounding the house forms the pleasure ground while the other half is a woodland and trial ground. There is hardly any formal structure and the emphasis is very much on exuberant informality, with different environments gradually merging into one another. The juxtapositions of these environments are cleverly handled with changes of levels and turf mounds, as well as good solid walls.

Lawns give way to rock gardens, woodland to trial garden in a lazy, unhurried way. The garden is one in which to leisurely walk, with only the occasional imaginative flight of fancy to interrupt one's reveries. Such fancy is nowhere more evident than in the rock garden which abuts Bracken Hill House. Like the house, the rock garden is a strange creation with walk-through arches and little, contrived pools which descend to a serpentine, if somewhat truncated, stream. Among the towering rocks are set masses of ferns but the real joy is to be gained by negotiating the little paths and crossing the hump-back bridge across the pool.

At the back of the house, the eastern side, a terraced lawn drops to a short stream scooped from a gulley, the sides of which are held in check by further outcrops of rock. In the water *Caltha palustris* grows, its vivid, bright flowers like giant buttercups. At the

water's edge hostas form dense weed-suppressing clumps and a small willow, *Salix gracilistyla* sits like a contorted bonsai. Caucasian speedwell runs wild and celandine flowers early in the spring. An *Acer palmatum* 'Senkaki' adds a splash of red with its bright and vulgar young branches, while its mature bark is a more subdued and refined pink. Beyond the stream and bordering the road is a long flower bed packed with musk thistles, teasels, giant sea holly and woolly-headed thistle.

Three tall mature conifers – *Sequoiadendron giganteum*, *Cedrus atlantica* and *Cedrus deodara* – dominate the eastern end of the house. Nearby another rock garden describes a curvilinear eastern boundary to the garden. Many of these rocks have disappeared beneath a thick blanket of periwinkle, others provide homes for the most delightful range of saxifrages and stone crops while still more play host to a family of euphorbias. There are many craggy corners lush with arum and polypody while the numerous identification labels stand up like miniature tombstones. Close to the house a bed is given over to that most vigorous genus of plant, *Polygonum*. They are among the first plants to break the ground in early spring.

The front of the house faces south into the main body of the garden. The lawn here is dominated by a mature *Cedrus atlantica*, *Liriodendron tulipifera* and, slightly to the north an *Ailanthus altissima*, named the tree of heaven for its elegant appearance when mature. Around its feet grow thick clumps of *Arum italicum* and, in spring, sky-blue wood anemone and buttercup-yellow celandine.

A grassy mound hides the stable block from view. To the south a long border is given over to 100 species of *Hebe* at the end of which a short flight of steps climbs to the sunken garden, the only remotely formal area in this garden. It is here that the influence of Lutyens can be felt. A long narrow pool is surrounded by lawn and then raised flower beds. Many British native plants grow here: violets, wood anemones, meadow saffron, lungwort, woundwort – a long list of ancient evocative names. There are lush clumps of celandine, too, a plant which has dropped down from its raised bed and now studs the grass with its charming flowers.

Beyond the sunken garden is the woodland and to the right, the beginning of the eighty-yard (seventy-two-metre)-long herbaceous border. Many of the plants growing here provide seeds which form the nucleus of the garden's annual seed distribution list: *Lysimachia*, *Salvia superba*, *Iris sibirica*, *Tradescantia*, *Lythrum*. The herbaceous border dog-legs away to the right. To the left is the delightful woodland environment where

Bristol University Botanic Gardens: the emphasis is on exuberant informality.

The sunken garden.

many British native trees and shrubs grow. Among the sycamores and yews, ivy weaves a dense carpet.

Many of the more choice woodland plants are grown around a triangle of formal lawn just west of the woodland. *Trillium grandiflorum*, gentians, *Colchicum*, hellebores and *Meconopsis betonicifolia*, with its heavenly blue flowers, provide a seductive splash of colour. Within this sunny spot can be found *Liquidambar styraciflua*, a real oddity with its bark raised up in long brittle flutes.

A long tall wall defines the western extremity of the garden. The adjacent border contains groups of plants related either genetically or by sharing a common purpose or quality. Poisonous British plants, such as *Digitalis*, *Atropa belladonna*, and *Aconitum*, are grouped together. There is a broad belt of arums, then a mass of culinary herbs: tansy, fennel, chives, hyssop and sorrel, all of which exude delightful fragrances when bruised.

Between this border and the glasshouses is an acre (.4 hectare) of land cut by straight paths into large rectangular beds. Bare in winter, this area assumes a summer livery

Bristol: a place in which to leisurely stroll.

which changes with the years, and stepped brick terraces provide a home for hundreds of pot grown alpines. Further terraces have small pools to accommodate aquatics. One corner is covered by a huge fruit cage containing small trees while the main body of the garden is hidden from view by walls and a dense shrub border.

Beyond the wall are the greenhouses, the best of which is a lean-to containing epiphytes on tree trunks and orchids in flower almost every month of the year. Climbers such as *Plumbago capensis* and *Passiflora antioquiensis* scramble up twine and a young banana grows vigorously in a pit. Nearby is *Xanthosoma violaceum*, its beautiful arrow-shaped leaves carried on long purple-black stems which flare from the crown of the plant. There is also a bird-of-paradise flower, *Strelitzia reginae*, with its astonishing blooms of orange sepals and blue petals.

Alongside the greenhouses is a magnificent nursery display of young plants. From these the Friends of the Garden can choose ten annually. The garden also provides a seed list of over 500 specimens which is distributed to botanic gardens around the world.

Cambridge University Botanic Garden

Cory Lodge · Bateman Street · Cambridge · CB2 1JS
Tel. 0223-336265

Open Mon–Sat 8 am–6.30 pm
(summer),
8 am–dusk (winter);
Sun 2.30–6.30 pm May–Sept.
No entrance fee.

Economics have played a major role in determining the development of the Cambridge University Botanic Garden. Founded in 1760 on a five-acre (two-hectare) plot close to King's College Chapel, the botanic garden moved to its current forty-acre (sixteen-hectare) site in 1846. Although plans were drawn up for the whole area, only the western half of the garden was actually laid out, the other half being let annually as allotments. This arrangement lasted until 1950 when an injection of funds in the form of a bequest brought about rapid development of the remaining ground. The garden then found itself in the fortunate position of having space in which to experiment.

Although the history of the botanic garden is modern, the history of botany at Cambridge stretches back to the mid-seventeenth century and specifically to John Ray. Ray, the son of a blacksmith, entered the university in 1644 at the age of seventeen and went on to hold various teaching posts over the next eighteen years. In 1658 he made the first of his carefully documented botanical tours through the Midlands and the North of Wales. Two years later he published his *Cambridge Flora*, which listed 626 species in alphabetical order.

In the preface Ray sums up the contemporary state of botanical knowledge and lays an academic trace which he was to follow with great enthusiasm. 'I became inspired with a passion for Botany,' he wrote, 'and I conceived a burning desire to become proficient in that study, from which I promised much innocent pleasure to soothe my solitude. I searched through the University, looking everywhere for someone to act as my teacher and my guide . . . But, to my astonishment, among so many masters of learning and luminaries of letters I found not a single person who was deeply versed in Botany . . . What was I to do? Should I allow the flame of my enthusiasm to be quenched or diverted

The lake at Cambridge, the garden's most picturesque feature.

Left *John Ray, whose enthusiasm established a passion for botany at Cambridge University.*
Right *John Stevens Henslow, who took over the garden in 1825.*

to some other field of study? I decided that this must not happen. Why should not I, endowed with ample leisure, if not with great ability, try to remedy this deficiency?'

Ray's infectious enthusiasm soon established within the University a passion for botany and a tradition of what was to become known as 'herborising'; the colloquial term for the jaunts into the surrounding countryside undertaken by Ray and his friends.

The names of Gerard and George London, a gardener and nurseryman who subsequently went on to work at Chatsworth and Blenheim, have both been linked with early attempts to found a botanic garden at Cambridge – Gerard in 1588 and London over one hundred years later, when the then Vice-chancellor encouraged him to visit Cambridge in connection with a physic garden. Nothing resulted from the visit and it was not until 1760 when Richard Walker, Vice Master of Trinity College, purchased the five-acre (two-hectare) plot bordering Free School Lane that the idea of a botanic garden became a reality.

Walker had a very definite idea of a botanic garden's function. He saw its role as the study of both tender and hardy plants so that their virtues could be discovered for the benefit of mankind. His botanic garden was therefore conceived as a traditional physic garden. Any flowers and fruits produced were of secondary importance, a passing diversion en route to greater discoveries.

This view prevailed until 1825 when the garden was taken over by John Stevens Henslow, a twenty-nine-year-old botanist. Henslow revived botany's flagging fortunes at Cambridge and was to go on to number Charles Darwin among his pupils. He also had a natural passion for trees and it was immediately obvious to him that there was little innovation he could bring to a five-acre (two-hectare) garden that had been under cultivation for sixty years. He therefore lobbied for a larger botanic garden. In his address to the Members of the University of Cambridge on the botanic garden he defined clearly what he saw as the main role of a modern botanic garden.

'It may very safely be asserted,' he wrote, 'that the larger number of living species that are cultivated in a Botanic Garden, the greater will be the facilities afforded to us all; not merely for systematic improvement, but for anatomical and other experimental researches essential to the progress of general physiology. It is impossible to predict what particular species may safely be dispensed with in such establishments, without risking some loss of opportunity which that very species might have offered to a competent investigator, at that exact moment he most needed it.'

Henslow constructed a strong defence of botanic gardens and one that is as relevant today as it was then, and went on to suggest that the botanic garden could be a useful reserve for trees. In this way he hoped to justify in the eyes of the University the need for a much larger site. He wrote, 'The reason why a Botanic Garden requires so much larger space than formerly, is chiefly owing to the vastly increased number of trees and shrubs that have been introduced within the last half century.' Henslow's lobbying proved to be successful and it is his garden that survives today. At a stroke he created for the city of Cambridge not only a centre of scientific excellence but a public amenity of great beauty. The first annual report of the Management Syndicate in 1856 testifies to the value of the botanic garden as a social amenity. 'In the formation of the new Garden, it was intended to make science the first consideration; but now in order to encourage a general taste for botanical studies, and to render the Garden an agreeable acquisition to the University, the designers consulted ornamental appearance wherever it did not interfere with the main object'.

Henslow's planting of trees was further developed by Richard Irvine Lynch, who was appointed Curator of the garden in 1879. Lynch, who trained at Kew, established the international reputation of the garden by exchanging seed lists with other botanical gardens around the world. He also redesigned the water area and rock garden, and established the ornamental bamboo collection which still catches the eye of the visitor entering from the Brookside Gate.

Effectively the Cambridge University Botanic Garden is two gardens. The western half was developed when the garden was founded in 1846; the eastern half comprises the area occupied by the allotments let on annual tenancies from 1867 to 1950. In 1950 the Cory bequest allowed further development.

The forty-acre (sixteen-hectare), rectilinear site lies on an approximate east-west axis. Originally situated on a meadow some one and a quarter miles (two kilometres) from the city centre, it is now surrounded on all sides by low-rise Victorian buildings, many of which are used as student accommodation and English language schools. An old water course, Hobson's Conduit, constructed in the seventeenth century to bring chalk spring water from the Gog Magog hills, forms the garden's western boundary. A small appendage to the garden sits like the head of a hammer at the north-west corner.

This small contained square of lawn and trees faces the beautiful bow-back Victorian house purchased shortly after the Second World War which is now the Garden Office. From the lawn there are glimpses of the larger garden beyond, as well as examples of a number of striking plants. The aristocrat of hydrangeas, *H. quercifolia*, shelters by a wall, while in one bed grows an extravagant display of Gertrude Jekyll's most favoured plant, *Yucca gloriosa*, whose somewhat exotic structure seems slightly out of place in British gardens. A ribbon of lavender dissects the lawn while to one side is a great clump of *Sarcococca hookeriana*, a plant whose fragrance in late winter stops visitors in their tracks. The whole is closed off from the rest of the garden by a large *Cedrus atlantica* 'Glauca'.

From here the old western half of the garden opens up and Henslow's brilliant planting of trees can best be appreciated, disguising as it does the flatness of the site and creating sweeping vistas which are alternately closed or turned by mature beech and oak.

The old western garden is quartered by two paths. The Main Walk running on an approximate east-west axis is itself cut by the gently serpentine Henslow Walk. At the western end of the Main Walk stands the Trumpington Gate, an ornate wrought-iron gateway which came from the garden's original site in 1909. Alongside this gateway grows a magnificent common lime, *Tilia* x *europaea*, planted to commemorate the opening ceremony in 1846.

The path from the Brookside Gate skirts a thicket of Caucasian wingnuts, *Pterocarya fraxinifolia*, and crosses a small stream which runs from Hobson's Conduit and feeds the small horseshoe-shaped lake. In spring, growing beside this stream on the roots of willows, can be found the parasite, *Lathraea clandestina*, its large tubular purple flowers forming spreading clumps. Astilbes grow here in profusion and in late summer the startling blood-red stems of *Lobelia cardinalis* and the fiery spikes of *L. speciosa* add a splash of defiant colour.

Just west of the lake is a small woodland, its dark and peaty floor home for many primulas, hostas and euphorbias and the beautiful Welsh poppy, *Meconopsis cambrica*. The common cuckoo pint thrives there, its white cowled flowers followed by bright red berries on erect green stems. The rarest tree in the garden is here too, *Tetracentron sinense*, eighty years old and still quite healthy. Between the lake and the woodland is Lynch's original planting of bamboo. The lake is the garden's most picturesque feature. It is a haven for wildfowl, which nestle among the delightful waterside planting or float with great aplomb on the still water. The rush-like *Acorus calamus* 'Variegatus' grows there, its strap-shaped foliage like blades of a sword with its cutting edge flushed bright yellow. The lake is surrounded by trees with rich autumn colouring. Within the curve of the lake is a raised bank sheltering a boggy water garden, where the greater spearwort, royal fern and recently discovered great fen ragwort grow. On the raised bank can be found the small, sinuous half-hardy shrub, *Arctostaphylos manzanita*, with its grey foliage and bark the colour of polished mahogany.

Between the lake and Henslow Walk is the raised limestone rock garden, a feature grafted on to the nineteenth century landscape during the development of the 1950s. The rocks support plants from four continents: Asia, Australia, Africa and North America. The plants hug the outcrops and form mounds and carpets of subtle colour. The pasque flower, *Pulsatilla vulgaris*, may be beautiful in spring and the exquisite grey-leaved *Tanacetum densum* at other times, but they are nothing compared to the *Catalpa* x *erubescens* 'Purpurea', which breaks into flower in late summer, its heavy sprawling limbs propped for safety among the rocks. Each erect panicle is formed from upward of fifty tiny orchid-like flowers, swamping the surrounding terrain with a rich fragrance.

To the south beyond the lake and intersection of the axial paths, where stands that most humbling of trees, the wellingtonia, lies the true heart of the botanic garden. The systematic beds, a mass of curvilinear informal island beds, display eighty families of mostly herbaceous plants. From these beds comes the teaching material used in the practical classes in the University. For the visitor absorbed with the diversity of plant structure and form there are rich metaphorical pickings to be had there.

Diagonally opposite the systematic beds are the nine glasshouses: temperate, conservatory, alpine, stove, palm, aquarium, tropical fern, orchid and succulent. The present range was built in 1888 by Lynch and subsequently rebuilt in 1932. Each of the small south-facing glasshouses opens onto a common corridor backed by a high wall. The bays between the houses create micro-climates where a number of exotics, such as *Trachycarpus fortunei* and the pink and white forms of *Crinum moorei*, grow in bloated contentment. In nearby beds are tender annuals: love-in-a-mist, *Nigella hispanica* and the yellow horned poppy, *Glaucium flavum*, set off by lush blue-grey rue, *Ruta graveolens* and the pink tree mallow.

The corridor linking the greenhouses is hung for most of the year with exotic climbers, including a spectacular bougainvillea. Bird-of-paradise-flower, *Strelitzia reginae*, is here, too, its flowers held like threatening crested birds of prey, and *Hibiscus rosa-sinensis*, with its crushed-silk pink flowers.

The stove house is hot and clammy, popular on chilly autumn days but too small to transport the imaginative visitor to tropical climes. African violets nestle beneath the tender *Mimosa pudica* whose leaves are so sensitive that they fold when touched. A cool contrast is provided by the alpine house. Its plants, whose origins range from the Balearic Isles to North Africa, are set in raised plunge beds, while dozens of tiny saxifrages are tucked into pockets in the rocks in the central landscaped beds. The palm house is the largest of the glasshouses. Its central lantern reaches forty-five feet (fourteen metres) but even this is insufficient. The tallest tree is a heavy-leaved fig, and members of the

Acanthaceae family thrive in the low light levels below on the forest floor. Economic plants, such as tea, banana and rubber, allow visitors to relate the commercially available substances to the original sources.

By far the most interesting glasshouse, the carnivorous house, stands by itself just beyond the main range. It is a small structure, the glass obscured by green algae. Here can be found the North American pitcher plant, *Sarracenia*, which tempts flies into its deadly watery brew by a subtle combination of colour and fragrance. The British native sundew, *Drosera rotundifolia*, employs a different yet equally effective method of dispatching its victims. Its rosettes of leaves are covered in sticky red hairs which trap the insects and then digest them.

John Parkinson writing in the 1640 edition of his *Theatrum Botanicum* offers an evocative description of the sundew: '. . . It hath divers small round hollow leaves, somewhat greenish, but full of certaine red haires that make them seeme red, every one standing upon its owne footstalke reddish hairy likewise, the leaves have this wonderfull propertie that they are continually moist in the hottest day, yea the hotter the Sunne shineth on them the moister they are.'

The Dutch, Parkinson informs us, call the plant lustwort.[1] But we must turn to Gerard, writing many years earlier, for this earthy explanation: 'Because sheepe and cattell, if they do but only taste of it are provoked to lust.'[2]

Distilled water from the sundew, '. . . a glittering yellow colour like gold', according to Gerard, when taken in wine with 'Cinnamon, Cloves, Maces, Ginger, Nutmegs, Sugar, and a few graines of Muske,' is claimed to ease consumption. But to take the brew neat is to court disaster.

Gerard says that, '. . . it hath also been observed, that they have sooner perished that used the distilled water hereof . . .' – a warning that we should always follow the physician's directions. The sundew's most deadly property of all, administered to the hapless fly, seems to have been missed by both Gerard and Parkinson.

A large area of the eastern half of the garden is given over to private research, and the expanses of lawn are dominated by some fine trees. There is a tall, elegant tree of heaven, *Ailanthus altissima*, from China and a *Paulownia tomentosa*. This Chinese native is known as the foxglove tree. The buds of its spectacular flowers are produced in autumn and overwinter on the tree, so are vulnerable to frost. The leaves are like massive green fans. Other notable trees include cherry, birch and alder, together creating a variety of shifting interest. At the eastern extremity are clustered the island beds of herbaceous plants, providing huge accents of bold colour between spring and autumn.

Off the main path, and forming a crescent which drops down to a shelter before rising again, is the scented garden, where sweetly scented shrubs, flowers and herbs, create an intoxicating cocktail of summer fragrance.

Just a short step away to the west is the chronological border, the garden's most deliberately didactic feature. Planted by John Gilmour, this long narrow bed chronicles

The winter garden, an area of bold colour in a subdued season.

the introduction to Britain of many of the most familiar herbaceous and shrubby plants. Among those introduced before 1550 are rue and acanthus. Rue was called Herbe grace by the old herbals for its '. . . many good properties whereunto it serveth; for without doubt it is a most wholesome herbe', according to Parkinson. It was served, he says, as an antidote to, '. . . venomous things, as well as Serpents,' and as a contraceptive, destroying '. . . the ability of getting children'.[3] The chronological border speeds the visitor across the centuries: echinops, 1550–70; snow-in-summer, 1631–50; yucca, before 1691; right up to the present day with the tiny *Geranium dalmaticum*.

Beyond the chronological border, but just off the main path, is the winter garden, an area of bold colour in the drab season. Hellebores, conifers, heathers and *Viburnum* x *bodnantense* create the effect of an Impressionist painting. The fresh young growth of crimson and yellow dogwoods form dense thickets after being repeatedly cut back. The bizarre blackberry, *Rubus biflorus*, is a mass of arching silver-white stems set with razor-sharp thorns.

1. John Parkinson, *Theatrum Botanicum*, 1640.
2. John Gerard, *The Herball*, 1636.
3. John Parkinson, ibid.

Above *Chelsea Physic Garden circa 1850 from the river, before the construction of Chelsea Embankment.*

Below A *copperplate engraving from Lyson's* Environs of London *(1795) showing the garden from the river.*

The Chelsea Physic Garden

66 Royal Hospital Road · London · SW3
Tel. 01-352 5646

The fortunes of the Chelsea Physic Garden have fluctuated considerably since 1673 when it was conceived by the Society of Apothecaries as a garden in which to grow medicinal plants. But it has survived and is now the second oldest botanical garden in England. The four-acre (1.6-hectare) site has remained in continuous cultivation throughout the period but has remained essentially a private teaching and research garden financed originally by the Apothecaries and, more recently, by a body of Trustees.

For most of the time the public have been rigorously excluded. Recently, however, the garden's exclusivity has been breached. The Trustees have decided that the garden must become self-financing, and to that end the public are admitted on two afternoons each week during the summer.

This is not enough to solve the problem. The maximum number of visitors the garden can accommodate annually before damage is done to the paths and grass does not bring in enough money. Therefore additional income is being sought by inviting external institutions to make use of the garden's facilities, such as the herbarium and library, and to hold educational courses within the garden itself. Thus the garden's historic role of education remains intact.

When the Apothecaries chose the site in 1673 Chelsea was a village surrounded by fields. Its proximity to the Thames allowed easy access by barge as well as ensuring that the garden enjoyed a constant high water table. The Apothecaries saw the garden as a living textbook where students could learn to recognise plants and to become familiar with their medicinal and scientific properties. At the time such information was generally only available from three sources; herbals (often notoriously unreliable), herbariums and the pursuit of 'herborising', going out into the surrounding countryside to identify plants in situ.

Many early herbals were based on the doctrine of 'like cures like'. Even early editions of Gerard's *Herball* had to be treated with a degree of caution until, in 1636, Thomas Johnson published a substantially enlarged and corrected edition.[1]

Herborising was a pursuit made popular in Cambridge by John Ray in the 1650s (see page 36) and was certainly practised by Johnson and the Society of Apothecaries. That method, however, tended to limit first-hand knowledge of plants to indigenous species. To acquire familiarity with the large number of exotics entering the country necessitated having a garden in which to grow the seeds in controlled conditions. The Apothecaries therefore leased the ground from Charles Cheyne and almost immediately erected a high wall around the four acres (1.6 hectares).

The early years were difficult. The garden put an immediate financial strain on the Society, whose fortunes and property had already been reduced in the Great Fire of 1666. Although some members suggested that the scheme should be abandoned, most were in favour of keeping it going. Money was raised among members and within ten years of the garden's foundation, when it was under the care of John Watts, himself an Apothecary, it attracted the attention of the Professor of Botany at Leiden University, Paul Hermann. In 1683 Watts reciprocated the visit and brought back with him the famed *Cedrus libani*, four of which were planted in the garden. These trees were among the earliest cedars of Lebanon to be planted in England. Although they were regarded as something of a novelty, the trees grew so well that two eventually had to be cut down in 1771 because they created too much shade. The remaining two trees survived until 1878 and 1904 respectively.

The garden had two other visitors of note during those early years: Doctor, later, Sir, Hans Sloane, and John Evelyn, both of whom have left written accounts. Sloane, President of the College of Physicians and of the Royal Society, was destined to secure the long-term future of the garden. He bought the Manor of Chelsea from Cheyne in 1712 and ten years later leased the ground to the Apothecaries in perpetuity, providing it was always used as a physic, or what we would now call a botanic, garden.

Sloane studied at the garden during his early training as a physician and left a record of a 1684 visit in a letter to John Ray.[2] 'I was the other day at Chelsea, and find that the artifices used by Mr Watts have been very effectual for the Preservation of his Plants, insomuch that this severe winter has scarce kill'd any of his fine Plants. One thing I much wonder to see, the Cedrus Montis Libani, the Inhabitant of a very different climate, should thrive here so well, as without Pot or Green House to be able to propagate itself by Layers this Spring.'

The following year John Evelyn, the great seventeenth-century arboriculturist and author of *Sylva*, also visited the garden and was equally impressed by Watts's skills and his greenhouses. 'I went to see Mr Watts, keeper of the Apothecaries' Garden of simples at Chelsea, where there is a collection of innumerable rarities of that sort; particularly, besides many rare annuals, the tree bearing Jesuit's bark, which had done such wonders in Quartan Agues – what was very ingenious was the subterranean heate, conveyed by a stove under the conservatory, all vaulted with brick, so as he has the doors and windows open in the hardest frosts, secluding only the snow.'[3]

Watts was obviously something of a pioneer but the good times at Chelsea were short lived. Watts seemed to lose interest in the garden and it drifted into one of its cyclical troughs which lasted until 1722 when the Apothecaries appealed to Sloane for help.

As well as leasing the garden to the Apothecaries, Sloane was instrumental in securing the appointment of Philip Miller as Curator, a tenure that was to last almost fifty years, during which time the garden's international stature was established. With Miller the golden age had arrived.

Within two years of coming to Chelsea Miller published his much acclaimed *Dictionary of Gardening*, in which he demonstrated his accurate observation of plants. So famous did the garden become under his care that the great Swedish botanist, Linnaeus, who refined and developed the binominal system of plant classification, visited in 1733.

By all accounts the meeting between the two men was difficult, with each adopting a hostile posture. Miller pointed out the plants in the garden to Linnaeus using the old-fashioned, long-winded Latin descriptive names. Linnaeus remained silent and unenthusiastic. On their second tour of the garden, Linnaeus indicated the same plants and named them according to his simplified system. Miller at first resisted the new simplified nomenclature but eventually succumbed to its benefits. Indeed, later editions of his *Dictionary* (which went to eight editions) adopted the Linnaean system. There is one other notable achievement for which Miller will be remembered. In 1732 he sent cotton seeds to the new American colony of Georgia. From those seeds the greater part of today's American cotton is descended.

By the middle of the century the gardens were considered among the finest in Europe. Peter Kalm, a pupil of Linnaeus, visited in 1748 and wrote, 'We saw Chelsea Hortum Botanicum, which is one the principal ones in Europe. (It is) one of the largest collections of all rare foreign plants, so that it is said in that respect to rival the Botanic Gardens of both Paris and Leyden'.

Miller had transformed the garden's fortunes and firmly set its course for the future. In the closing years of his reign he even received 500 packets of seed from Sir Joseph Banks, who had collected them on his round-the-world voyage with Captain Cook on *Endeavour*. Banks, naturalist, founder of the Horticultural Society, horticultural advisor to George III and, as such, the first unofficial director of Kew Gardens, had great affection for Chelsea having, as a lad, lived close by at Swan Walk.

Miller barely outlived Banks's return in 1771. After forty-eight years at the garden he was pensioned off and died the following year aged eighty. He was succeeded by William Forsyth (after whom the forsythia is named). The following year Banks, ever the explorer, was in Iceland. He brought back with him slabs of lava cut from the lava beds of Hecla and presented them to the physic garden. The lava was joined with stone from the Tower of London to make what is thought to be the earliest rock garden in the country. In Victorian times a pool was created on top of this rock garden and the whole feature turned into something of an oddity.

William Curtis, another Apothecary who went on to publish in 1787 his famous

A profusion of summer colour in the order beds.

Botanic Magazine, joined the staff of the garden. Many others associated with the garden are remembered through plants. One, Nathanial Ward, Master of the Society of Apothecaries, invented in 1863 the Wardian case, a sealed glass container which was instrumental in moving tender plants across continents and seas without injury. This simple device has proved popular ever since; it was used by Robert Fortune, the plant hunter and curator of the Chelsea Physic Garden from 1846–8, to transport 20,000 tea plants from Shanghai to the Himalayas. An original Wardian case can still be seen in the fern house at Chelsea Physic Garden today; inside the case grows that most rare of ferns, the Killarney or bristle fern, *Trichomanes speciosum*.

By 1897 the garden had become a real financial burden to the Society, and the Treasury instituted an enquiry into its future. The outcome was the transfer of the garden to the City Parochial Foundation to be administered by a body of Trustees. This situation remained until 1981 when the Trustees decided that if the garden was to survive it must become self-financing. At last the gates were thrown open and the general public admitted.

Although the delineation of the beds has changed over the years, the garden retains its

Autumn colour in the Physic Garden.

rigid formality. It is in effect a quartered square with the axis occupied by a replica of Rysbrach's statue of Sir Hans Sloane, the original having found a home in the British Museum. The cruciform shape of the garden, however, is progressively drowned during the summer by the profuse planting. The order beds which occupy almost a third of the garden become foaming masses of green, and the woodland corner, great banks of solid verdure. More than one hundred plant families are spread out upon the ground and delightful relationships can be found. In one bed the foxglove tree, *Paulownia tomentosa*, grows amid a clump of *Digitalis*.

The woodland floor at the north-east corner of the garden possesses its own special charm. *Trillium grandiflorum*, with its waxen pure-white flowers, carpets the ground before the bold leaves of hostas unfurl. *Smilacina racemosa*, Solomon's seal and kirengeshoma occupy the filtered shade beneath a mighty holm oak, *Quercus ilex*, and a leggy yew, *Taxus baccata*.

Around the perimeter of the garden dense thickets of shrubs and trees have been allowed to grow in order to obscure surrounding buildings and to give the impression that the garden is larger than it actually is.

Dense thickets of shrubs and trees give the impression Chelsea Physic Garden is larger than it really is.

In the borders flanking the embankment are hypericums, hardy ferns and thick, lush clumps of paeonies. In spring a tall cherry, *Prunus* x *yedoensis*, drops a perpetual shower of blossom and the tangled tall white branches of *Rubus cockburnianus* present a curious, eye-catching web of silver. A sunken pool is surrounded by beds of salvia and various bergenias, the best of which is *B. ciliata* with its dazzling blooms. Another mighty holm oak extends the woodland gloom into the main body of the garden while a tall tree of heaven, *Ailanthus altissima*, adds an elegant counterpoint.

Flanking the central paths is a winter woodland garden, bright with the yellow-green flowers of *Helleborus corsicus* and white, bird-like blooms of *Magnolia soulangiana*. Winter brilliance is extended by *Daphne tangutica* and a wide belt of *Sarcoccoca confusa*, whose discreet flowers give off a delightful honey-like fragrance. In early spring *Magnolia stellata* produces its starry white, fragrant flowers; later, Martagon lilies and the almost black-flowered *Geranium phaeum* take over.

On the west border is the fern house. Packed with ferns in landscaped beds, this greenhouse has a Victorian fussiness, but it is worth a visit if only to see that original Wardian case, situated at the south end.

Beyond the fern house and running along the west boundary of the garden is a bed of Australasian plants. Eucalyptus, hebes and senecio grow here beneath a tall silver wattle, *Acacia dealbata*, its yellow flowers scenting the air in late winter. At the end of this bed in the north-west corner of the garden, a formal parterre has been created with brick paviors. These beds are still being developed but will eventually house a selection of plants introduced into Britain through Chelsea during Philip Miller's long tenure.

The northern boundary of the garden is occupied by the garden buildings, none of which is of particular merit. The plants which grow in the protection of the south-facing wall do deserve attention, especially *Fremontia californica* with its brilliant buttercup-yellow flowers carried on soft downy branches from May until October.

In long beds facing the garden buildings is the national collection of *Cistus*, commonly known as rock rose, while to the east *Helleborus corsicus*, lily-of-the-valley, *Viburnum* 'Anne Russell', and a brick red *Chaenomeles*, make a beautiful combination beneath the delicate spreading branches of a flowering cherry.

Sadly, the glasshouses along the north wall have reached a dangerous condition and now await replacement. In front of them is an ancient gnarled cork oak, *Quercus suber*, and a wonderful bed of sun-loving Mediterranean plants.

Beyond the ageing olive tree, medicinal plants and herbs in beds edged with fragrant santolina are ten dwarf box mounds which create a charming miniature avenue.

The overall effect at Chelsea is one of profusion within a contained area where plants are grown for scientific rather than aesthetic reasons. The garden is a plantsman's garden, one not of views but of plants and the story that goes with them. It currently distributes an annual seed list to 350 botanic gardens throughout the world and this, along with the garden's emphasis on education, amenity and conservation, typifies the modern role of a botanic garden.

1. Thomas Johnson, botanist and herbalist, exhibited in his shop window in Snow Hill in London in 1633, the first bunch of bananas ever seen in Britain. He first hung them in April and they survived until June when he sliced and ate them. His description in Gerard's *Herball* of 1636 is worth quoting at length for its historical interest: '. . . my much honoured friend Dr Argent (now President of the College of Physicians of London) gave me a plant he received from the Bermuda's. The fruit which I received was not ripe, but greene, each of them was about the bignesse of a Bean; they all hang their heads downwards, they somewhat resemble a boat: the huske is as thick as a Bean, and will easily shell off it: the pulp is white and soft . . . The Stalke with the fruit theron I hanged up in my shop, where it became ripe about the beginning of May, and lasted untill June; the pulp or meat was very soft and tender, and it did eat somewhat like a Muske-Melon'. Johnson, an ardent Royalist during the English Civil War, was killed in 1644 defending Basing Castle in Hampshire against Cromwell's troops. The aforementioned bunch of bananas were immortalized when they were engraved for the frontispiece of Johnson's edition of Gerard's *Herball*.
2. Ray's Philosophical Letters, as quoted in *The Romance of the Apothecaries' Garden*, p. 35, Cambridge University Press, 1928.
3. Ibid.

Cruickshank Botanic Garden

University of Aberdeen · St Machar Drive · Aberdeen · Scotland · AB9 2UD
Tel. 0224-480241 ext. 5247

Open Mon–Fri 9 am–4.30 pm;
May–Sept, Sat–Sun, 2–5 pm.
No entrance fee.

The Cruickshank Botanic Garden occupies eleven acres (4.45 hectares) of ground in Old Aberdeen some way from the granite grey heart of the city. It lies hidden behind a clutch of university buildings, wedged between St Machar Drive and The Chanonry. In this modestly elegant Victorian road is the garden's main entrance, itself a modestly proportioned wrought-iron gateway. Through the gateway and beyond the clutch of buildings the garden opens out.

The original square plot dates from 1898 when Miss Anne Cruickshank established the garden with the dual aim of serving both the educational needs of the university and the public as an amenity. Ten years later the grounds of number 8 The Chanonry were acquired and the garden extended beyond its immediate and original boundary. The greenhouses and rock garden stand on this new plot, and a further acquisition of land allowed the development of an arboretum just to the north and linked to the main body of the garden by a footpath. Mature trees now play a major part in the landscaping of the garden, breaking the skyline and creating many private and secluded areas.

The oldest part of the garden is a large, almost square plot with a sunken garden, systematic beds and long double-sided herbaceous border. Many old favourites can be found in this border: phlox, Japanese anemone, achillea, daisies, *Eremurus robustus* – a good eight feet (2.4 metres) in height – as well as some unusual species. There is a finely cut acanthus with gorgeous pink veining along the centre of each leaf, and the holy thistle, *Silybum marianum*, its glossy dark-green leaves delightfully marbled. Myth has it that this marbling occurred when milk spilled from the breast of the Virgin Mary as she suckled the infant Christ.

The border is terminated by a great swag of *Geranium macrorrhizum*, its leaves deliciously pungent when crushed or bruised.

The sunken garden is a curious affair, a great bowl scooped out of the earth and obscured by the mass of shrubs and conifers which grow on its slopes. Serpentine paths run down between carpets of heather and dwarf rhododendrons such as *R. campylogynum*, itself sheltered by more bold and vigorous species such as *R. augustinii*. In spring

The long double-sided herbaceous border.

bulbs grow in profusion accompanied by the glaucous leaves of *Iris pallida*. In late summer, summer hyacinths, *Galtonia candicans*, raise their stately heads and pale-coloured *Colchicum* species stud the grassy floor of the bowl. *Polygonum amplexicaule*, that hard-working clump former which flowers from June until September grows there, and *Euphorbia robbiae* spreads in the light shade close to *Rodgersia podophylla*, which in turn creates dramatic leafy accents with its bold-as-brass foliage.

Beyond the sunken garden, the systematic beds radiate like the spokes of a wheel. Sheets of *Salvia horminum* sparkle in the sun, their upper tissue-paper bracts flushed pink, purple and mauve. *Lythrum salicaria* sends up its fiery plumes well into September, while the sinister, poisonous *Aconitum japonicum* cohabits with *Helleborus corsicus* and a belt of *Nigella damascena*.

The aconitum has long been recognized as an extremely poisonous plant. In his *Herball* of 1636 Gerard describes, with characteristic vigour and detail, the lethal effect of swallowing just a few leaves; '. . . when the leaves hereof were by certaine ignorant persons served up in sallads, all that did eat thereof were presently taken with most cruell symptoms, and so died. The symptoms that follow those that doe eat of these deadly Herbs are these; their lipps and tongue swell forthwith, their eyes hang out, their thighes are stiffe, and their wits are taken from them.'

Of the aconite's British cousin John Parkinson wrote in his *Theatrum Botanicum* of 1640, '. . . the hunters of wilde beasts, doe use to dippe the heads of their arrowes they shoote, or darts they throw at the wilde beasts, which killeth them that are wounded therewith speedily.'

A high red brick wall separates the original garden from the later 1909 acquisition. Alongside this wall but still in the old garden is a small patio area made in 1980, its flagstones and dry walls now all but overrun with plants. The patio is littered with pots and ancient troughs. Gypsophila sprays out a great inflorescence alongside two magnificent stone urns, each of which contains variegated lemon-scented pelargonium, and the dry wall is host to some delightful sedums.

The wall is pierced by a tiny gateway which gives entry to the 1909 garden. Here the land ascends gently. To the left is a small woodland area dominated by old oaks. Hellebores grow here among the woodland flora as do hostas, Solomon's seal, epimedium and *Omphalodes cappadocica*. On the slopes the rock garden rises in small terraces and island beds punctuate the lawn.

Backing onto the wall which divides the old and new gardens is a mixed border. Among this collection of shrubs and perennials is the late flowering hawkweed from Germany, *Hieracium maculatum*, its leaves slashed with deep magenta-black stains. The aristocratic *Kirengeshoma palmata* is also found here, with its black stems and perplexing hood-shaped, waxy yellow flowers. The border runs down almost to the greenhouses, which are not open to the public.

Dundee University Botanic Garden

516 Perth Road · Dundee · Scotland · DD2 1LW
Tel. 0382-66939

Open Mar–Oct, 10 am–
4.30 pm;
Sept–June, Mon–Sat inc.,
July and Aug, Mon–Sat inc.
and Sun 2–5 pm.
No entrance fee.

Dundee University Botanic Garden is a young garden established in 1971, and therefore still under development. About twenty-two acres (8.9 hectares) in size, it occupies an elongated site – three rectangular oblongs none of which shares a common axis – on a gentle south-facing slope almost on the banks of the Firth of Tay. Although well landscaped, the garden's main emphasis is on education. There is a very good visitors' centre where displays are changed regularly.

At the heart of the garden is an ecological area which traces natural plant associations from mountain-top to coastal habitats. This area has the advantage of a water course fed by an on-site well and has attracted a great deal of attention from botanists. It offers the visitor a kaleidoscope of terrains which in the wild would otherwise almost certainly prove inaccessible. One can climb alongside the stream, ascending from lush waterside planting through deciduous woodland of oak, ash and birch where in spring celandines, primroses and bluebells grow. Adjacent is a conifer plantation, beyond which juniper scrub opens out. Many interesting ferns, such as the shuttlecock, *Matteuccia struthiopteris* and the scaly male fern, *Dryopteris felix mas*, grow by the stream, and there are also some fine clumps of common polypody, *Polypodium vulgare*.

As the conifers thin out, heathers become more prominent, interspersed with the bilberry, *Vaccinium myrtillus*, which grows well on thin, barren soil. The fruit of the bilberry, or whortleberry, when dark and ripe is deliciously thirst quenching. Of its other properties, Gerard proclaims in his 1636 *Herball*, 'They be good for an hot stomacke, they mitigate and allay the heate of hot burning agues, they stop the belly, stay vomiting'. In a delightful coda, he also asserts that the people of Cheshire 'do eat the blacke Whortles in creame and milke'. On slightly higher ground beyond the whortleberry is gorse, *Ulex europaeus*, smothered in buttercup-yellow flowers and guarded by vicious sharp spines. The stream is in a shallow gulley here with coarse scree leading up to the montane. From this vantage point fine views can be had of the Tay Bridge; amazingly, the climb from wet fen to dry scree covers little more than fifty-four yards (fifty metres).

Around this central part of the garden are various groupings of trees and shrubs, with conifers playing an important role. On the south border are cedars, spruces and pines, with some particularly fine specimens of Korean fir, *Abies koreana*. This handsome tree has ripe bluey-mauve cones which exude a sticky resin until they shed their peripheral seeds. The central cores remain standing upright like extinguished candles on the branch.

Birches and evergreen oaks, beeches and rowans also grow in this central area, although they are still quite small, as one would expect in such a young garden. But the quick-growing conifers more than compensate for what the garden lacks in age, as they grow in a way that allows their overall structure to be taken in at a glance.

At the eastern extremity development is still very much under way. There is a fine group of eucalyptus and how marvellous they will look in a decade when their ghostly silver limbs bend over the surrounding foliage. There, too, is the beautiful *Acer pensylvanicum*, perhaps the most visually exciting of all the snake-bark maples to come from North America. Its bark is exquisitely striped with silver green and tawny; its leaves, fresh and bright in early spring, become drab in late summer only to emerge in autumn in rich nut-brown finery.

At the western end of the garden, close by the entrance, are the visitors' centre and glasshouses. Sheltered behind yew hedges is a neat formal herb garden where culinary and medicinal plants sun themselves and grey-leaved Mediterranean exiles hope for warmer days. Although comparatively small, the temperate and tropical glasshouses contain much interesting material and are similarly landscaped to those at Edinburgh.

In a watery basin the giant water lily, *Victoria cruziana*, unfurls its large plate-size leaves, one every six days or so. Its diurnal flowers are fragrant on a warm summer's evening. Close by is a wonderful exotic from the New World Tropics, *Xanthosoma violaceum*. Its large heart-shaped leaves are dusky green on top, flushed with purple and delicately veined underneath, and spring on bluey-purple stems directly from the ground. Tangled above in a silvery grey web is Spanish moss.

In the desert section the cactus *Cereus peruvianus* flowers late in the year, its large yellow blooms attached to the succulent stems like an afterthought. The Brazilian shrub, *Tibouchina urvilleana*, flowers magnificently there too, with wide-eyed purple blooms a good three inches (7.5 centimetres) across, and *Bougainvillea glabra* scrambles among the roof supports, showering down a cloud of purple flowers.

Giant water lilies in the tropical glasshouses at Dundee.

Durham University Botanical Garden

University of Durham · Hollingside Lane · Off South Road · Durham
Tel. 0385-64971

<table>
<tr><td>Open every day,
9 am–4 pm.
No entrance fee.</td></tr>
</table>

Normally, the sight of a monkey puzzle tree crammed into the front garden of a gloomy Victorian villa scarcely merits a passing glance. At the University of Durham Botanic Garden, a single specimen of *Araucaria araucana* stops you dead in your tracks. The tree is magnificently displayed on the crest of a low rise of land in the very heart of the garden, its curlicue branches making a dramatic silhouette against the sky. When viewed from across a nearby pond, it is enhanced by the light-green feathery foliage of four young dawn redwood, *Metasequoia glyptostroboides*. The effect is quite extraordinary.

Founded in 1970 as a low-maintenance garden, most of Durham's eighteen acres (seven hectares) is laid to turf. A considerable range of trees is grown as individual specimens, which gives the garden the feel of a public park. There is, however, more to Durham than this.

The site is a south-facing slope shaped like a chunky L with the entrance at the tip of the foot, the highest point of the garden. Just inside the gates is a coarse scree bed, with scree and boulders which are still rather raw and therefore eye-catching; the overall effect is vaguely reminiscent of a building site. From there the garden slopes gently away. On the west flank a serpentine path follows the contours of a long shrub border and set in the gently sloping lawns is an island herbaceous bed, a splash of vivid colour with purple loosestrife, *Lythrum salicaria*; and *Echinops ritro* evident in late summer.

The dog-leg of the garden is obscured by a reasonably dense planting of North American conifers. There the slope becomes steeper, dropping to an extensive area of undulating lawn. Young specimens of Lodgepole pine, Californian redwood and Douglas fir are much in evidence. At one point the undulations open into a tiny valley, the grassy sides of which are quite precipitous.

The serpentine path continues along the valley floor until it meets farmland and turns east. This is the southern extremity of the garden and the gulley separating farmland from garden is thick with cow parsley, foxgloves and willowherb.

The path follows the valley through part of an old woodland. Climbing up through this woodland another path, flanked by a rustic fence, zigzags through clumps of holly

Above and below *Summer and autumn views of Durham University Botanical Garden.*

and beech until it deposits the visitor among a thicket of rhododendron. Lost among the trees a stone slab keeps alive the memory of one young soldier who died during the Great War: 'P.J.R. In France. 17 October 1915' is inscribed upon the grey stone surface. This is also the home of the common stinkhorn fungus, *Phallus impudicus*, with its evil stench.

At this point there are several surprises. The pond is suddenly revealed, with the path running across the dam and on to a long swag of hostas. From this point, the monkey puzzle is perfectly framed and the water's surface presents a mirror image of the scene while beyond a mini-parkland unfolds.

If the monkey puzzle dominates the view west, then that to the east is equally dominated by a tall silver birch. A number of interesting shrubs shelter there; particularly arresting are *Kolkwitzia amabilis*, with its tiny delicate foxglove-like flowers growing from wine-red stems, and *Cornus kousa*, with its white bracts tipped pink and looking like the finest hand-made silk.

The water from the pond is channelled across a wide slate slab. The water runs slowly, just enough to create movement and wrinkle the pond's surface before dropping into a shallow rill. Beyond this rill is a round stone plateau reached by steps. Surrounding the plateau is the relatively new heather garden whose slopes should be completely carpeted

Torch ginger (Phaeomerica magnifica).

within the next few years. Above it, foliage effects predominate; variegated weigela, *W. florida variegata*, flutters in front of *Cotinus coggygria* which in turn highlights an evergreen privet, *Ligustrum lucidum*, and a mock orange, *Philadelphus coronarius*.

Beyond this dazzling display, lawns run up to a magnificent yet modest academic residence. This house is separated from the main body of the garden by a ha-ha, an effective barrier from which spring holly saplings, foxgloves and the distinguished biennial Aaron's Rod, *Verbascum thapsus*, its magnificent stalk encrusted with small yellow flowers.

A short way up the slope toward the garden entrance, the glasshouses and rose garden occupy a plateau. Only the arid and the tropical glasshouses are open to the public. Both are nicely landscaped and both are packed with plants. The largest plant in the tropical house is the pawpaw, *Carica papaya*, its upper branches seeking liberation through the roof. Another contender for the tallest plant is a yellow-stemmed bamboo, its smooth surface marked with fine green stripes with such precision that they could have been painted by the finest calligrapher.

Although young and by no means spectacular, the garden at Durham is interesting for its striking topography. In one respect it is extraordinary: the city of Durham is dominated by the cathedral, yet from no point within the garden's boundaries is it visible.

Dendrobium thyrsiflorum, *growing in Durham tropical glasshouse.*

Fletcher Moss
Botanical Gardens

Millgate Lane · Didsbury · Manchester · M20 8SD
Tel. 061-434 1877

> Open every day,
> 7.45 am–dusk.
> No entrance fee.

The Fletcher Moss Botanical Garden in Manchester, part of the larger Fletcher Moss public park, is hardly a botanic garden at all in the accepted definition of the term. There is no research area or herbarium tucked away behind the scenes and there are no students to be found foraging on hands and knees among the plant collection. The most striking feature is the large rock garden made on a quite precipitous south-facing slope. This, along with more recent attempts to create a wild flower area on an acre (.4 hectare) of adjacent land, does however make a visit worth while.

The Fletcher Moss Park was presented to the city of Manchester by the late Alderman Fletcher Moss between 1914 and 1919. In all, the football pitches and tennis courts cover about 21 acres (8.5 hectares) while the botanic garden itself is but a fraction of this.

The slope from which the rock garden is made drops from the terrace which leads from the Wilmslow Road. Narrow paths zig-zag among the rocks, while on a plateau on the upper level is a small alpine house. This cedar house, although potentially alluring, contains little of real interest, with basically a few lewisias doing their best to enliven the environment. *Lewisia columbiana* 'Wallowensis', with its tiny white flowers veined with pink, is the most charming. There is also a deliciously fragrant rhododendron growing in a large pot, its blooms not unlike regale lilies although less robust. *Lewisia tweedyi*, with blooms the colour of beaten butter, also caught my eye.

Outside the alpine house and running along the top edge of the garden is a deep bed of conifers where fastigiate and prostrate forms compete for attention. At the far end of the bed a *Picea pungens glauca* assaults the eye with its grey-blue elegance, while self-sown Welsh poppies, *Meconopsis cambricas*, do their best to brighten up the gloomy spaces between the trees and a very young *Metasequoia glyptostroboides* struggles to gain height. On the very edge of the slope Japanese maples flitter their finely cut foliage in the breeze. *Acer palmatum* 'Atropurpureum' is there, sheltering beneath its branches that most attractive of British natives, *Pulsatilla pratensis*.

From there the slope opens out. To the west grow Chusan palms, *Trachycarpus fortunei*, protected from savage winds by a high red brick wall which forms the boundary

Pond and stream at Fletcher Moss, sheltered by the garden's brick wall boundary.

of the garden. Across the rock at one's feet, yellow gold dust, *Alyssum saxatile*, and moss phlox, *Phlox subulata*, form dense vivid carpets of brilliant colour beneath trees such as *Cornus nuttallii*. This dogwood is an amazing sight in spring, with its almost insignificant flowers surrounded by upward-facing cream-coloured bracts.

Also sheltering in the lee of the wall is a tall *Garrya elliptica*, its grey pendulous catkins and grey leathery leaves a welcome and haunting sight. Nearby, *Rosa hugonis* clings to the wall and flowers early with small wide-eyed yellow blooms. Its immediate neighbours are *Ceanothus* 'Gloire de Versailles' and *Actinidia kolomikta*, the latter too young yet to have adopted the distinctive splashes of pink on the tips of the leaves.

There is also a bed of plants introduced by Manchester born plant-hunter Frank Kingdom Ward and donated by the Ness Botanic Garden. *Lilium mackliniae*, *Primula ioessa* and *Rhododendron macabeanum* all seem perfectly happy on this peaty slope.

Not far away Welsh poppies push their beautiful blooms through a huge spread of violet-coloured *Phlox douglasii*. A bushy Korean fir, *Abies koreana* stands there, its young cones carried like upright candles, and a lush green *Picea glauca* 'Albertiana conica' squats, compact and cone shaped among the rocks. Water gushes suddenly from a hidden artificial spring, and foams and gurgles between slabs of granite. More *Alyssum saxatile* and *Phlox subulata* repeat the yellow and violet colour theme and a species of low growing pink, *Dianthus*, adds a counterpoint with its grey foliage. The water drops to two irregular shaped ponds linked by a serpentine stream.

The planting is profuse. Mature clumps of New Zealand flax, *Phormium tenax*, thrust up their strap-shaped leaves, hostas and rodgersias form weed-suppressing swaths and ornamental rhubarb, *Rheum palmatum*, spreads its vast leaves over the ground. The aristocrat of the group is undoubtedly the royal fern, *Osmunda regalis*, an ancient clump of which grows with its toes in the water, its crown thick and woody and its fronds tall and stately. There is also marsh marigold, *Caltha palustris*; skunk cabbage, *Lysichiton americanus*; and bistort, *Polygonum bistorta*, the vigorous and dense mat-forming plant usually considered a weed but worth space for its early light pink blooms, if there is room enough. The American cowslip, *Dodecatheon meadia*, is there too, its petals reflexed and swept back, as is *Primula helodoxa*, its whorls of yellow flowers carried on tall stems. Variegated hostas catch the eye next to Solomon's seal, and the delicate flowering inflorescences of sweet cicely add a delicate honey fragrance to the air.

Beyond all this, thickets of rhododendrons flourish, spangled with their bright vulgar blooms, and a tall *Liriodendron tulipifera* flashes its curious-shaped leaves. The flowers of the latter, which open in June, earn the plant its common name, tulip tree. Nearby is a tall leggy yew and copse of pencil-thin incense cedar, *Libocedrus decurrens*. Sheltering in the clearly defined spaces between the trees are regale lilies and the dark blades of *Phormium tenax* 'Purpureum'.

The rock garden here embraces a low scree slope where *Dianthus*, *Saxifraga*, *Pulsatilla* and *Androsace* grow, before petering out into a low heather garden. A *Viburnum carlesii* floods the air with a heady fragrance from its waxy white flowers and a low dome of jasmine box, *Phillyrea angustifolia*, seems a cross between common box and privet.

The wild flower area is still very much under development. Its long grass is studded with narcissus and cowslips, and clumps of various conifers lend a variety of shape to the scene. There also two clumps of vigorous *Reynoutria japonica* are quickly gaining ground and will have to be severely dealt with in a year or two.

Close by is a small enclosed garden surrounding an old parsonage. There can be found an orchid house where cymbidium flower in late winter and early spring and an old mulberry tree among a border of lavender.

Glasgow Botanic Gardens

Great Western Road · Glasgow · Scotland · G12 0UE
Tel. 041-334 2422

Open every day:
grounds 7 am–dusk;
Kibble Palace 10 am–4.45 pm
(winter),
10 am–4.45 pm (summer);
main range Mon–Sat 1–4.15
pm (winter),
1–4.45 pm (summer),
Sun opens 12 noon.
No entrance fee.

The botanic gardens in Glasgow stand conveniently close to the home of Scottish television and an agreeable symbiosis exists between the two. The gardens contain one of the most beautiful examples of Victorian architecture, the Kibble Palace, a curvilinear conservatory of iron and glass. Its interior is often used as a backdrop for television interviews, and its magical spatial conceits form the location for many television dramas. In 1985 visitors to the gardens mingled with actors dressed in eighteenth century costumes as a play on the life of George Frederick Handel was enacted.

Thus are the eighteenth and twentieth centuries happily combined. In reality, however, the botanic gardens are a product of the nineteenth century, a period when so many similar ventures were being undertaken to accommodate the huge number of exotic plants flooding into Britain from abroad. Almost every major city was raising subscriptions to build a botanic garden.

Glasgow's original version was laid out in 1817 on an eight-acre (3.2 hectare) site at Sandyford in Sauchiehall Street. Before then, a physic garden had existed for almost one hundred years in the grounds of the old College. The driving force behind the new botanic garden was Thomas Hopkirk, a local botanist of some note who, in 1813, had published his *Flora Glottiana*, a catalogue of 662 species found close to Glasgow. The same year he issued a catalogue of the many plants growing in his own private garden at Dalbeth. This collection was to form the core of the new gardens.

Funds were raised through private subscription. Hopkirk played an active role by lobbying hard among his friends. The University donated £2,000 and, in return, a room was allocated in the gardens in which the newly appointed Professor of Botany could deliver his lectures. The Royal Botanical Institution of Glasgow, formed by Hopkirk to turn his dream of a botanic garden into a reality, undertook to supply specimens to the University which, in 1818, established its first Regius Chair of Botany. The first Professor was Dr Robert Graham who laid out the grounds with the help of Stewart Murray, the gardens' first curator.

In 1820 the great botanist William Jackson Hooker was appointed to the Chair of Botany at Glasgow. As a botanist he displayed consummate skill and enthusiasm, but as a teacher and administrator he could not have been a worse choice. During his tenure the gardens prospered and the collection grew rapidly. Between 1821–5 the number of species in the gardens grew from 9,000 to 12,000, many of them newly introduced exotics. Gathered together under Hooker's judicious eye, these plants helped establish the gardens' international reputation.

The rapid expansion of the collection brought problems, however. It became increasingly obvious to Hooker that the land at Sandyford was too small. By the time he left Glasgow in 1841 to become Director of the Royal Botanic Gardens at Kew, plans had been made to purchase a larger site just a short distance to the north at Kelvinside.

The collection was moved and the present garden was opened in April, 1842. The annual subscription to members was one guinea but the general public could gain admittance on Saturdays for the payment of one shilling. Occasionally the charge was reduced to one penny to give the working classes a chance to have a look.

At Glasgow the Institution clearly saw one of the gardens' roles as being that of a public amenity. However, they were slow to realize that to maintain the garden to the level desired by members, it was necessary to provide additional attractions to tempt paying customers.

In 1871 the Institution invited John Kibble, engineer and entrepreneur, to move his extraordinary glass palace to the gardens from his estate at Coulport, Loch Long. In a complex contract drawn up between him and the Institution, Kibble was allowed to create what can only be described as a Victorian fun palace. The Institution agreed to pay him an annual subscription – an outgoing that was to dog the gardens for many years and which was to eventually propel the Institution into parlous economic problems – while any profits were to be shared. The palace was dismantled and moved by raft up the Clyde to Glasgow where it was enlarged on re-erection.

For Kibble, moving the palace from his estate at Coulport to Glasgow was a sound commercial venture. The airy, shimmering extravagance of the palace was an immense attraction in its own right while in addition he intended to hold concerts in its central chamber. The conceit of having a band playing tucked out of sight beneath a pond in the central chamber was to be the crowning glory. The estimated profits from such marvels would be Kibble's with only a portion allotted to the gardens. But the risks were Kibble's too. For the right to hold concerts the contract stipulated that he had to maintain the structure for twenty-one years.

The deal proved disastrous. Kibble eventually lost interest in the venture and the burden on the Institution of the annual payment precipitated a crisis. In 1881, with the aid of a loan from the Corporation of Glasgow, Kibble's lease was bought out for £10,000, but it was too late. Dwindling subscriptions and the fact that the Institution had over a number of years overreached itself forced them to seek a more permanent solution.

Statue of Eve inside the magnificent Kibble Palace in Glasgow.

Glasgow Corporation took over the gardens in 1887 and after more than forty years' continuous expansion, the botanic gardens closed. Fortunately, the setback was only temporary.

Four years later, after the City of Glasgow Act had been passed, the gardens were reopened. The City undertook to maintain them as a public park while at the same time continuing the relationship with the University. The obligation to supply plant material remained and remains today, a requirement which has prevented the gardens from becoming merely another public park.

The term public park perhaps loosely sums up a visitors' first impression of the gardens. The broadly rectilinear site aligned on a south-west axis spreads away, rising gently to the north and west. To the north is the River Kelvin, to the south the Great West Road. But a great deal of work takes place out of public view, and for the visitor prepared to dig deeper behind the scenes, the gardens' quietly pervasive scientific role can still be found.

The gardens have many obvious didactic elements. A chronological border and systematic beds are most obvious, but the best is the herb garden laid out in 1957 and modelled on old monastic herb gardens. The more obvious didacticism, however, does seem half-hearted, secondary to the more elementary role of creating a pleasing environment. This is no bad thing and the environment is pleasant enough, with a vast number of trees and recently developed arboretum to the north. If the visitor feels disappointed with the gardens' general disposition, the feeling is soon dissipated by the range of glasshouses. Like the Kibble Palace, these have attracted world wide fame.

The Kibble Palace is the most exciting and perhaps the most perfectly formed glass and iron structure in the country, out-ranking in breadth of vision and audacity of execution even Kew's elegant palm house. The palace is the heart of the gardens and it is difficult to find a spot where it does not dominate the view.

Eric Curtis, the gardens' current curator, believes that Kibble's architects were Boucher and Cousland, local men who were responsible for many of Glasgow's villas, some of which had small ornate conservatories attached. The palace is a conglomerate of buildings, the largest of which is round in plan and appears at first like a low squashed dome. It looks like a perfect aerodynamic form, a fact which Eric Curtis believes explains the palace's ability to withstand storms and gales.

The interior splendours of the central chamber are only gradually revealed via a curvilinear glass tunnel running from the axis of additional glass wings and beneath a modest dome. The glass orifice falls away and the vegetation billows up, a mass of ferns pierced through by tall old Australian tree ferns, *Dicksonia*, their dark trunks a network of fibrous roots and their uppermost crowns splayed out beneath the shimmering glass dome like huge Egyptian hieroglyphs. Set coyly among this tangled growth are alabaster figures, some nude yet all discreetly placed around the central flanking path. On the periphery is a collection of Victorian camellia cultivars which flower in late winter. The building has a breathtaking luminosity and a translucence difficult to imagine. It is the epitome of architectural elegance, where a finely tuned balance is achieved between space and light.

It is impossible not to be moved by the grandeur of the tree ferns seemingly trapped beneath a web of glass and iron. There is a fragility and yet a permanence to the scene. The Kibble Palace re-creates paradise within a glass sphere; everything within is ordered and contrived to lift the eye and the heart. Even the cast-iron columns spiral up to Corinthian capitals where sit the lightest of filigree iron brackets.

It is Renaissance arrogance tempered by Victorian originality. There plants from the temperate regions of the world grow alongside the exotics beloved by the Victorians. There is a tiny patch of moss unique to this glasshouse and new to science when discovered on site in 1920. There too is Crusoe's fern (*Thrysopteris elegans*) waist high and yet somehow inconspicuous and now almost extinct on its native Pacific island of Juan Fernandez. This fern was only recently added to Kew's threatened plant list and

Sir William Jackson Hooker, whose work did much to establish the Glasgow Botanic Gardens.

attempts are being made at Glasgow to propagate it, although no fertile spores have been found on the plant in eighty years.

Both Gladstone and Disraeli spoke within Kibble's eloquent creation, but even they would not have heard the orchestra as was originally planned, because Kibble's dream of music being played out of sight beneath the pond was never realized. Today the pond is no more and the central arena is thick with ferns.

The other main range of glasshouses – eleven in all – occupy higher ground just north of the palace. They are altogether more modest in style and do not attempt to compete with the palace in any way. They range from the succulent house, where in January 1983 the bromeliad *Puya alpestris* flowered its peacock-blue blooms for the first time after November gales stripped glass from the roof, to the palm house, a melancholy yet strangely satisfying space where a Benjamin fig, *Ficus benjamina*, stands tall among elegant palms propped up on long fibrous aerial roots.

The National Begonia Collection is found in this range and the species are among the most extensive in the world. In the old begonia house is Eric Curtis's special treasure, *Begonia foliosa*. Planted directly into the soil, this plant is never without at least one flower, a record that stretches back over ten years.

Other curiosities include the St Helena Ebony, a species thought extinct until a plant was found on the island in 1980. The seeds were brought back, Kew germinated them and now Glasgow, only one of three botanic gardens in the world to possess the plant, has two seedlings.

In the aquatic house, growing in a circular raised tank, is the giant Victoria water lily. Always a favourite with children, this extraordinary plant is raised annually from seed, its huge leaves being produced every six days during July and August. Sadly, the gardens' other gem, the filmy fern house, is not open to the general public. In the cool, damp conditions of this dank room grow ferns so delicate and fragile that their fronds make the finest tissue paper seem like coarse linen.

Harlow Car Gardens

Crag Lane · Harrogate · North Yorkshire · HG3 1QB
Tel. 0423-65418

Open every day,
9 am–7.30 pm or sunset if
earlier.
No entrance fee.

In 1948 the Northern Horticultural Society leased forty acres (sixteen hectares) of land from Harrogate Corporation. Their intention was purely didactic, to create a garden where members of the Society could see the results of trials which tested how various plants adapted to the often unfavourable conditions in this cool and damp corner of England. Of the original land just under half was nothing but fields. The rest was woodland, a remnant of the ancient Knaresborough Forest. Subsequently the acreage has swelled to seventy (twenty-eight hectares), the size of the garden today.

The remarkable achievement of the Society is plain to see. What would seem on first sight to be an unfavourable valley site has been turned into a garden of wide-ranging interest. It is difficult to believe that the garden is in fact the result of less than forty years' work; the garden looks as if it could have been there for a century or more.

The site occupies a valley on an approximate east-west axis through which runs a stream. There are sulphurous springs there too, the health-giving properties of which were not exploited until, in 1840, Henry Wright built a hotel and bath house over one of them. The bath house, a well proportioned single-storey building of warm honey-coloured sandstone, is now the garden's study centre and library.

Because the garden's obvious didacticism is aimed at domestic gardeners, there are many small areas to which the average gardener can easily relate. There are also trial areas where vegetables and flowers are rigorously and scientifically tested, and even an area for testing composting methods.

Curiously, it is the south-facing slope of the valley that is clad with birch and oak woods and a more open arboretum. The north-facing slope, once arable fields, now contains garden plants, and is cut by a number of descending beech and yew hedges

The extensive rock garden at Harlow Car.

Harlow Car Sulphur Spa and Hotel, 15 May, 1857, showing the bath house on the left.

which lend an overall geometry to the scene. (There is, however, no overall plan binding together the disparate parts and within the garden's obvious boundaries the planting has a slightly amorphous feel.) The tall hedges that drop down the slope along with the Broad Walk play a vital role in anchoring the meandering disorder of island beds and trial areas to the site, adding in the process a welcome sense of order and formality.

In gardens of this type the overt didacticism can be overwhelming. They require many visits repeated over a number of years. If time is short, dive straight into the garden and try to touch the elusive, *Draughtsman's Contract* spirit of the place. Harlow Car repays such treatment. When saturation levels have been reached on the open slopes one can cross the luxuriantly planted stream, and enter a world of trees. This stream separates two very different worlds, the contrived order of the open garden on the north-facing slope and the wild woodland opposite. Between the two is achieved an almost perfect balance, nature tamed played off against wilderness, albeit somewhat nurtured wilderness. Whether or not the metaphysical achievement is deliberate, those visitors with a romantic nature will be at home here.

Trees are an essential ingredient to the enjoyment of Harlow Car. Their presence can never be ignored for more than a moment and they continually take the eye upwards away from the fabulous display of planting on the ground. For, in spite of the lack of order, the planting here deserves considerable, if selective, attention.

The most eye-catching feature is the stream-side garden which is unmatched anywhere in the country and achieves variety and interest along its whole length before disappearing beneath the study centre. Shuttlecock ferns, hostas, royal ferns, primulas, foxgloves and Solomon's seal abound. The beautiful parasite, *Lathraea clandestina*, grows on the roots of willows; skunk cabbage, *Lysichiton americanus*, floods the air in early spring with its evil smell and *Iris sibirica* precedes a colourful display of candelabra primulas.

Above the stream's banks, maples, azaleas and conifers form the transition into the abutting woodland. A number of bridges lend additional interest and over the whole garden hangs the acrid, yet not altogether unpleasant smell of sulphur. This pungent tang is never more evident than by the study centre, where the capped sulphur wells leak.

Climbing above the study centre on the fringe of the oak woodland are peat terraces punctuated with huge rocky outcrops, which in the damp atmosphere are hosts to many emerald-green mosses. Ferns grow fat and rich in the cool humidity and acid soil: varieties of hart's tongue, *Asplenium scolopendrium*, with divided fronds; soft shield fern, *Polystichum setiferum divisilobum*; and the coarse yet beautiful hard fern, *Blechnum spicant*. In autumn the encroaching oaks and birches drip moisture; in spring the hardy orchid, *Dactylorrhiza elata*, lends its aristocratic presence to the scene and the ubiquitous native Welsh poppy, *Meconopsis cambrica*, seeds itself in the most unlikely places.

The Fellows Walk along the stream allows the visitor to tread the delicate balance between order and wilderness. To one side the woodland rises in inviting majesty, to the other the open garden unfolds.

Central to the unfolding open garden is the extensive sandstone rock garden. There masses of alpines and small bulbous plants, including daffodils and snowflakes, flower in spring while the autumn maples and slow-growing conifers produce a dignified tapestry of melancholy colour. Sandwort, *Arenaria balearica*, makes an invasive living green carpet, and fescue, *Festuca glauca*, diffidently tucks blue spikes among the rocks.

There are vivid gentians, iridescent scillas and miniature hebes. The rampant *Acaena inermis* from New Zealand has marched steadily over the rocks and invaded the narrow path so that it is now difficult to walk without treading on the coarse grey-brown foliage. Further up the slope and deeper into the rock garden conifers begin to grow tall. There are fewer alpines and space is given to great swathes of heather of the most subtle pinks and creams. A young Scots pine, *Pinus sylvestri*, its foliage like upturned sprays, towers above the bushy wiry form of spruce, *Picea abies* 'Nidiformis'. The luxuriance of planting in the rock garden is astonishing.

The rock garden joins the lily pool where a prostrate spruce, *Picea pungens* 'Koster' grows beneath a *Metasequoia glyptostroboides*, which was grown from one of the seeds sent back to Great Britain from China in 1941. There is also a delightful *Gladiolus papilio*, a delicate, almost fragile, plant flowering in late autumn. Its pink-mauve blooms are shot through with white and carried on arching stems. Above the pool stretches the

The dramatic Broad Walk at Harlow Car Gardens.

tarn meadow, completed in 1972, where heathers and conifers grow in island beds and from where, on clear days, the distant Harlow Moors can be seen.

Through a thicket to one side, the Broad Walk boldly descends the slope. The view is sealed at the bottom by the trees on the opposite slope. Prominent among these is an American beech, *Nothofagus obliqua*.

Away from the tarn meadow is an area devoted to bulbs. In spring thousands of daffodils can be found in island beds while later in the summer extravagant lilies flood the air with their scent. A tall conifer hedge separates the bulb meadows from the rose garden, which is rigidly geometric and awash with fragrance until Christmas. In front of the hedge is a border of annuals: nigella, cornflowers, godetia, viscaria and candytuft, in confectionery pinks, mauves, yellows and blues.

Few visitors penetrate deep into the woodland garden. This is a pity for there can be found an unrivalled forest silence. Within the birch wood the undergrowth has been allowed to grow naturally and there is a startling luminosity as the light flickers through birch foliage. Underfoot the ground is soft and springy with masses of celandine on all sides. The conditions are ideal for rhododendrons, which grow in great dense evergreen thickets.

The birch and oak woods are separated by a path, slashed transversely by an equally broad ride which gives excellent views along the whole length of woodland. There is a glade, where the translucent birch canopy gives way to the opacity of oak, and one is confronted by an extraordinary folly, the remains of a pseudo-classical portico standing on a platform. Two couchant lions, their stony heads held high and their features almost obliterated by wind and rain, guard the steps leading to the platform.

The portico is all that remains of the entrance to the Cheltenham Spa Rooms, a social hot spot at a time when Harrogate enjoyed great popularity as a spa. When the spa rooms were demolished in 1939, the portico was rescued and moved to its present site to become an 'eye catcher' within the woods. It is at its most striking in winter when the surrounding foliage has died and the tall grey columns, glimpsed from a distance, assume a poetic grandeur.

The oak wood is altogether gloomier than the birch wood. Light seems to be absorbed among the thicker canopy. There grows *Rhododendron rex*, its leaves, with their characteristic tan coloured undersides, perfectly matched to the soft light. Beyond the oak wood is an arboretum where some fine specimen trees grow. The Norway maple, *Acer platanoides* 'Schwedleri', creates an arresting spectacle at the foot of the slope. The maple's autumn colours are a vivid brilliant-orange next to Italian alder, *Alnus cordata*, a coarse-leaved tree with smooth grey bark. Near by is a hybrid service tree, *Sorbus* x *thuringiaca*, which has been brutally coppiced to a dense thicket of branches.

Above the arboretum on the open slope at the east end of the garden are the more obvious trial areas where vegetables are grown. The neat allotment-sized plots are watched over by a model of a sparrow hawk on a whip-like stave. Shrub roses edge a flagstone path which leads to a small foliage garden enclosed between hedges of yew and other conifers. Rue, santolina, *Juniperus squamata* 'Meyeri', and the heather-like *Cassinia fulvida* are just a few of the choice plants grown here for their beautiful foliage.

Beyond the alpine houses and patio area, where tiny didactic rock gardens litter the ground, the visitor comes to the other end of the trial garden where a small area is devoted to ground cover and hedging. Here in bays between hedges of symphoricarpos, hornbeam, hypericum and escallonia, a variety of succulent herbs prove their efficacy as ground cover plants.

Hull University Botanic Garden

Thwaite Street · Cottingham · Hull
Tel. 0482-849620

Open Thurs, 1–4 pm.
No entrance fee.

Only about a quarter of Hull University Botanic Garden is open to the public. Yet within that quarter can be found two quite different and equally appealing elements: the central corridor of the greenhouse complex located in the busy working half of the garden, and the long herb border, which shelters beneath a nearby high red brick wall. Both share a carefree appearance and a seemingly reckless disregard for aesthetics. The greenhouse corridor displays a sensuous profusion of bloated plants growing in large pots; the herb border has an air of abandon, with its subtle fragrances apparently thrown together at random.

In many respects, these two elements typify the working botanic garden that Hull definitely is. The visitor is as likely to encounter a white-coated student, punk hair gelled to perfection, as an elderly volunteer, also white coated, with trug full of freshly harvested seeds neatly packaged in brown envelopes.

Since the garden's creation in the early 1930s, it has grown outwards to embrace the grounds of adjacent houses and that of the nearby Thwaite Hall. The area open to the public is dissected by a row of mature trees; in one a *Vitis coignetiae* has climbed high into the crown. To one side stands the greenhouse complex, to the other, the grounds of a large Victorian villa now used as student accommodation.

The garden behind the villa is mainly lawn but there are island beds of plants grown in family groups, and old conifers positioned to emphasise their structure, colour and form. Prostrate, for example, is set against fastigiate, and deep greens against blues, contributing a sense of maturity to the scene.

The lawns sweep away in curves to create the erroneous impression that further attractions lie just out of sight, beyond the next bend. At the far end is a short curving border dominated by two *Thuya occidentalis*, growing so close that they form a tall, bulging teepee shape. This border is home for plants which are always a joy to see, even if they are not unusual – sea lavender, *Limonium latifolium*; bloody cranesbill, *Geranium sanguineum*; Siberian bugloss, *Brunnera macrophylla*; and the ghostly giant Scotch thistle, *Onopordum acanthium*.

Island beds of plants grown in family groups at Hull University Botanic Garden.

Nearby are the island beds where related plants, such as the kniphofias, hostas, lilies and hemerocallis of the *Liliaceae* family, grow.

The herb border, which is almost lost from view behind these beds, is a riot of epicurean fragrances and tangled growth. Rue, santolina, sage, lavender and mint, with their burnished Provençal reds, muted yellows and greys, are delightfully set off by the red brick wall towering above them. Nearby is an avenue of apple trees which in spring dazzle with blossom.

Returning to the working half of the garden, a Mount Etna broom, *Genista aetnensis*, draped in yellow, pea-like flowers, adds a brilliant sunny splash to the scene. The colour is echoed by a young *Catalpa bignonioides* 'Aurea', whose rich yellow foliage is set off against a hedge of trimmed conifers. There are some fine specimens of hydrangea, too; *H. sargentiana*, with its velvety leaves; *H. aspera macrophylla* and *H. serratta* 'Thunbergii', with its delicate pink flowers.

The long central corridor of the greenhouse presents a leafy tunnel in which exotics grow and fuchsias hang in baskets from the roof struts among swags of *Aristolochia brasiliensis* with its curious bulbous flowers.

There are impressive ferns such as the grove fern, *Alsophila australis*, set in a huge clay pot, and the Australian tree fern, *Dicksonia antarctica*, growing directly in the ground. Though striking, the ferns are almost overwhelmed by a dazzling display of *Hibiscus rosa-sinensis*, whose blood-red blooms are the size of a man's hand; and by the pendulous tresses of a large *Cupressus cashmeriana*.

At one end of this corridor grows a *Trachycarpus fortunei*, and the wide-spreading fronds of the palm, *Pritchardia Korlaea*, look as though they could have been stripped from an Egyptian hieroglyph.

Off this central corridor are the smaller temperate fern, warm temperate and cool greenhouses. In the temperate fern house is the stag's-horn fern, *Platycerium bifurcatum*, sitting on a low plinth like a satiated monarch. An altogether more attractive fern is the nearby polypody, *Polypodium subauriculatum*, growing at head height in a basket. Its lacy drooping fronds often reach a length of four feet (1.2 metres).

Hume's South London
Botanical Institute Botanic Garden

323 Norwood Road · London · SE24 9AQ
Tel. 01-674 5787

What is the common link between the Indian National Congress and the South London Botanical Institute? Both were founded by the same man, Allan Octavian Hume; the Congress in 1883 and the Institute in 1910. Today the South London Botanical Institute possesses the smallest botanic garden in the country and boasts a membership of some 200 people. Hume, it seems, was one of that fascinating breed of philanthropic Victorian gentlemen, a patriarchal figure with a keen interest in natural history and a belief that the pursuit of the subject was something from which all mankind could benefit.

As a youth Hume had acquired a vast natural history collection, mainly of bird-skins and bird's eggs, which he presented to the Natural History Museum in London in 1855. In its day it was the largest single collection offered to the museum.

Born in London in 1829, Hume first went to India at the age of twenty where he joined the Bengal Civil Service, being appointed to the North-West provinces. He rose gradually through the civil service until becoming secretary to the Government of India in 1870.

During his career he developed a real love for the Indian people. Increasingly, he felt that they should have a say in their own government and he was convinced that only through some form of parliamentary system could the mass be educated and their miserable economic conditions be done away with. To this end he lobbied the educated elite of the nation.

Eventually under Hume's guidance an association of prominent Hindus convoked the first session of the Indian National Congress in December 1885. However the association, far from being a transitional step towards the establishment of a parliamentary system, soon became a rallying centre for those who desired the overthrow of the British.

Parallel to his political activities Hume continued his interests in ornithology. When he left India in 1894 he was considered by many to be the 'pope of Indian ornithology'. At that time he also realized that he had achieved all that he could for the Indian people.

On his return to England he settled at Norwood in South London and set about seeing what he could do to improve the lot of the local workers and artisans. He was convinced that science was the best avenue through which that improvement could be procured.

But what branch of science? The question exercised his mind for some time. Geology seemed the ideal natural science for the city dweller to pursue; the study of it would promote sound mental activity and improvement of mind as well as encouraging the artisan out into the fresh air. However the local geological interest was rather limited and Hume soon realized that botany (a subject about which his own knowledge was rather scant) was a more practical alternative.

With true paternalism he set about founding the South London Botanical Institute, a task that was to take him ten years. In 1910 he bought the house in Norwood Road in which the Institute still meets today.

Having bought the house and its half-acre (.2 hectare) of grounds, Hume began acquiring a herbarium and library. Both are still very much in use. He died in 1912, having barely achieved his aim, but not before extracting from the president of the Institute, A. B. Rendle (who also held the post of Keeper of Botany at the British Museum in London) a promise that he would not let the new venture founder. In the seventy-five years since then the Institute has survived many vicissitudes. That it is a sound establishment today is largely due to the voluntary efforts of Frank Brightman, the current president.

Brightman, a bewhiskered and avuncular man, knows that nothing dies as quickly as a garden. In 1975, when he took over, the garden was by then reduced to a fifth of an acre (.1 hectare) and sadly neglected, overrun with sycamores and undergrowth. His first five years were an uphill struggle with bouts of vigorous activity interspersed with periods of doldrums.

Today the Institute thrives, with over 200 members on its books. There is a busy calendar of events throughout the year and the Local Educational Authority holds regular evening classes there.

Number 323 Norwood Road is a three-storey, double-fronted Victorian villa standing in a not very fashionable part of South London. In domestic terms it is a substantial building with a flight of stone steps up to the front door and a semi-circular drive off a busy and noisy main road.

If the garden has of necessity changed over the years since Hume's death, the house itself has remained unaltered. Time there has very much stood still. Beyond the front door is an Edwardian interior where the hall is dominated by a huge ornate carved wooden clock. Gas lamps, their curved stems like heron's necks, can still be found in odd corners.

To the right of the hall is the extensive library housing 3,000 volumes specific to the plants of the British Isles. Through the hallway and to the left is the herbarium where, in the semi-darkness, light seems to lurk rather than shine and the atmosphere is heavy and musty. More than 100,000 specimens are housed there, mounted between sheets of brittle card and stacked in black steel boxes, one on top of the other.

Real delights are to be found in this garden by those prepared to spend the time looking for them.

Beyond the herbarium and through the small conservatory is the garden, a small rectangular patch divided into order beds by flagstone paths. In high summer it is a place of wild, profuse growth while in winter it adopts the semi-derelict air familiar to any suburban garden. It can all be taken in at a glance. Two minutes, perhaps even less, and one could be round the garden and out of the gate by the far side of the house. But the real delights of this garden are to be found by those prepared to get down on their hands and knees to forage among the specimens.

There is a fine collection of bramble. One, *Rubus thibetanus*, with coarsely toothed leaves which are whitish underneath, is the most handsome foliage plant of this species. Its brilliant silver-coloured stems, however, are coated in vicious thorns. Another speciality is the genus *Polygonum*. Many species grow but annual species thrive, including *P. persicaria*, often considered to be nothing but a weed, and *P. hydropiper* with its burgundy red stems. The latter has recently attracted the attention of applied entomologists because its leaves have been found to contain a substance repellent to aphids.

Rare plants represented in the garden include *Salix rosacea*, whose only natural habitat on mainland Britain is on Cwm Idwal in North Wales; and an apple-flavoured rhubarb, which appeals to Frank Brightman because of its historic association with nearby Deptford, a market garden area in Victorian times. Nearby is a tiny clump of Asarabacca (*Asarum europaeum*), that most beautiful and discreet of plants flowering in June with blue blooms tucked well out of sight.

Rescue operations are also launched from the Institute. One such snatched the spotted orchid, *Dactylorhiza fuchsii*, from the jaws of a bulldozer when a chalk pit was being filled in near Snodland, Kent.

There is a small desert strip too, covered by a neat glass frame. The jojoba, *Simmondsia chinensis*, grows here as do the tiny rosettes of *Agave recurvata*, grown from seed sent from the Desert Botanical Garden in Phoenix, Arizona.

Another beauty which thrives is the rare four o'clock plant, *Mirabilis jalapa*. This shrubby plant, with its beautiful yellow, pink, white and red blooms, was the first ornamental to be introduced from Peru. Of it Gerard in his *Herball* of 1636 writes, 'The floures are very sweet and pleasant, resembling the Narcisse or white Daffodil, and are very suddenly fading; for at night they are floured wide open, and so continue untill eight of the clocke the next morning, at which time they begin to close (after the maner of Bindweed) especially if the weather is very hot: but the aire being temperat, they remain open the whole day, and are closed only at night, and so perish, one floure lasting but onely one day.'

The seeds of the four o'clock plant are easy to grow, and the tuberous roots of established plants can be stored in a frost-free place. Gerard again writes, '. . . notwithstanding it may be reserved in pots, and set in chambers and cellars that are warme, and so defended from the injurie of our cold climate.' Gerard was obviously smitten by this 'wonderfull herbe', as he called it, and so is Frank Brightman. It really is a marvellous plant. My own clump was parented by just one specimen given to me by that great English plantswoman, Rosemary Verey.

The Institute takes its botanical role very seriously. Specimens from the herbarium have been loaned to the University of Utrecht for research purposes. Seed lists and seeds are also exchanged with forty botanic gardens throughout the world, and their own list last year contained sixty-two items. It is a prodigious achievement and one that testifies to the enthusiasm of the Institute's members.

Leicester University Botanic Garden

Stoughton Drive South · Leicester
Tel. 0533-717725

The history of the University of Leicester Botanic Garden dates from 1920, when a small garden was planted in the grounds of the then Leicestershire and Rutland College. The garden comprised a woodland glade, medicinal plants, herbs, a rock and water feature, and a systematic area for British native plants. In 1947, the University needed to expand their buildings and the garden was moved the short distance south to Oadby. There, over the next twenty years, it was to expand into a mature site of some sixteen acres (six hectares). Not all this land was immediately available. At that time much of it formed the grounds of four magnificent mansions built for wealthy local industrialists just after the turn of the century.

Three of these houses – Beaumont, Southemeade and Hastings – were designed by the architect, Stockdale Harrison, the earliest, Hastings, built in 1902. Beaumont was built next and its garden, which employed ten gardeners when at its best and achieved a considerable reputation, was laid out in 1905. Southemeade, more sombre and austere than the others, was a much later addition, built in 1928 as a retirement home for the then owner of Beaumont. The fourth house, The Knoll, was built in 1907 by a local builder, William Winterton, for his own occupation. It was acquired by the University in 1964.

All but Southemeade have a Lutyens feel about them, with delightful red bricks, external timbering, tall Elizabethan chimney stacks and stone mullions. The Knoll, infused with more imagination than the other three, even had specially made Tudor bricks and was roofed with Swithland slates. Today all four houses have an aura of fading gentility about them, but with their individual integrity remarkably intact. They are now used as halls of residence by the students of the University, which is just a bicycle ride away.

Three of the houses are fronted by a flagstone terrace with steps leading to terraced lawns surrounded by shrub borders. The exception is Southemeade, where the immediate environment is confined to a rather brutally paved herb garden. There is throughout great emphasis on trees; many of these have now reached their prime, contributing a

The Sandstone Garden looking towards Beaumont.

unique blend of woodland and helping to create what can only be thought of as a 'Gardenesque' environment.

John Claudius Loudon had helped define the term 'Gardenesque' when in 1838 he had written in his much acclaimed book, *The Suburban Gardener and Villa Companion*, that the style was 'calculated to display the individual beauty of trees, shrubs and plants in a state of nature; the smoothness and greenness of lawns and the smooth surfaces, curved directions, dryness and firmness of gravel walk, in short, it is calculated for displaying the art of the gardener'.

Before the integration of all four gardens at Leicester, Loudon's theories were shoe-horned into areas of some four or five acres (1.6–2 hectares). The integration began in 1964, and has been achieved remarkably successfully, allowing each garden to retain some of its individual character. Hedges and terraces blur but do not obliterate the old boundaries so that the effect is one of a continuous change of emphasis within a structured whole. The total site forms the shape of an elongated axehead, falling gently to the south.

Beaumont is the most impressive area within the garden, with a number of disparate elements carefully combined to achieve a garden of some distinction and eloquence. The east and west lawns are dissected by a watery gulley where great slabs of sandstone disguise the change of levels. This gulley is a mass of ferns, hostas and *Helleborus corsicus*, with a narrow forest of maples on the sides. *Acer japonicum* 'Aureum', its leaves shimmering like beaten gold, and *Acer palmatum* grow to the west side while to the east *Acer palmatum heptalobum* 'Elegans' and *Acer japonicum* 'Aconitifolium' spread their leaves above a border of low growing plants, with a tight-knit patchwork effect. Prostrate conifers and cotoneasters, *Hebe* 'Carl Teschner' and a low growing abelia all continue the display well into the winter.

This emphasis on structure and form continues in the border below the terrace, which is best viewed from the east lawn. Hebes, *Ceratostigma plumbaginoides*, *Genista*, *Euonymous alatus*, the golden-stemmed *Cassinia fulvida* and many other special plants present a dazzling display. Above them, against the house, is a *Magnolia grandiflora*, its creamy-white blooms carried throughout the summer, while at the magnolia's feet butterflies bask on the richly scented sweet rocket, *Hesperis matronalis*.

The east lawn is edged by shrub borders above which tower the mature trees of the woodland walk. A mighty cedar, *Cedrus atlantica* 'Glauca' grows close by the house and this is balanced by the large round leaves of *Sorbus Mitchellii*, which are green on top, characteristically grey underneath.

Dominating this lawn to the south and serving as an 'eye-catcher' between an opening in the border, is a swamp cypress, *Taxodium distichum*. Beyond this tree, which turns a delicious foxy-red colour in the autumn, is a huge Norway maple, *Acer platanoides*. Among the dense planting rhododendron, pieris and azalea thrive.

To the east of Beaumont is a long herbaceous border backed by free-standing pillars linked by rope swags. These pillars, festooned and hung with roses, shield from view an attractive water garden around which the rose-covered pillars continue. Like so much else in this garden, the long rectangular pool, with an apse and fountain at the far end, has a Lutyens feel to it with perhaps just a touch of Gertrude Jekyll. There are watery recesses where rushes grow into thickets topped by chocolate-brown tapers.

The axis of the water garden is crossed by a pergola with arching, lichen-stained cross-beams, from which hang curtains of *Vitis davidii*. This vigorous vine, with its huge heart-shaped leaves and its ancient gnarled trunk, threatens to overwhelm the pergola. The vine is already locked in battle with the purple leaved *Vitis vinifera*, whose grapes are as hard as bullets.

Many other less invasive climbers scramble into the pergola: *Clematis* Marcel Moser, *C.* 'Lincoln Star', and *C. tangutica*, perhaps the most alluring of all, with its bearded fruits as soft as down. The base of the pillars is awash in sprays of ferns and hostas. There is a distinct Mediterranean profusion to this quiet corner of the garden, with its contrived dappled shade and cooling water. The feeling is reinforced and continued in the sunken garden tucked just beyond the pergola.

The sunken garden is a parterre with beds edged in dwarf box, nuttily fragrant in the sun, and an infill of variegated pelargoniums in late summer. The beds are separated by serpentine terracotta-coloured brick paths. Around the perimeter are yew hedges, while the sloping side terraces are alive with grey-leaved beauties such as artemisia, lavender, sage, salvia and santolina, which saturate the air with Mediterranean fragrances. This sunken suntrap is a beautiful place, a corner of repose set among dense verdure.

Above the sunken garden is a recently completed conservation garden where the beginnings of the National Collection of Bedding Violas are displayed in raised beds.

The transition from Beaumont to Southemeade and the rest of the garden is through a limestone rock garden where saxifrages and sempervivums grow amid weathered rock. Just to one side is the rose lawn, alive with scent for ten months of the year, and the National Skimmia collection, modestly sheltered beneath tall trees.

After the almost mutinous planting of Beaumont the herb garden surrounding Southemeade seems dull and in need of renovation. Even the thyme lawn dominated by a black walnut *Juglans nigra* and a red horse chestnut, *Aesculas* x *carnea*, is relatively unimpressive.

Further down the slope are the order beds, set in grass. No longer in use as a taxonomic resource, they have become a liability and their existence is threatened.

Hastings stands on a substantial terrace overlooking a fine lawn and grove of cedars, through which is reached a meadow. The border below the terrace is lush with *Nicotiana* and *Acanthus*, and an ancient *Wisteria sinensis*, thought to date from the original planting of the garden, covers almost all of the terrace's brickwork. The lawn is overshadowed by magnificent cedars, *Cedrus atlantica* 'Glauca', *C. libani* and *C. deodara*, among which grows the Japanese butterbur, *Petasites japonicus*.

The meadow, a delightfully unkempt area, contains nearly sixty British species of grass; in spring the discreet adder's tongue fern, *Ophioglossum vulgatum*, with its cowled tunic opening around a central spike, grows beside the curious yellow rattle, *Rhinanthus minor*. The whole area is dominated by a massive purple beech, *Fagus sylvatica purpurea*.

Beyond the meadow is the heather garden, where grass paths snake between large island beds. The most imposing of the garden's houses, The Knoll, can be glimpsed between tall conifers and a copse of silver birch. From the heather garden, terraced lawns lead up to The Knoll, set in a commanding position on a high terrace. Many of the garden's extensive collection of hollies are planted along its western boundary which gives way by Knoll Gate to a spinney of the native pine, *Pinus sylvestris*.

The garden's glasshouses are scattered. The largest, the main glasshouse, a modern, utilitarian structure, is sited on a terrace behind Hastings. The small alpine house is behind The Knoll and the ornamental, fern and succulent houses are ranged to the side of Beaumont.

The main glasshouse is divided internally into three sections, only two of which are open to the public. There are many exotics and a good display of economically useful plants, such as banana, rice, coffee and pineapple. Bulbs, orchids and insectivorous plants make a good display on benches, and an exotic touch is added by *Strelitzia reginae*, its orange flower poised like an insect ready to strike.

Recent restoration of the two adjoining alpine houses has allowed some permanent planting in raised beds with small cavities excavated in chunks of tufa rock to allow tiny rock plants to grow in the almost soilless pockets. *Corokia cotoneaster*, a low contorted angular shrub from New Zealand, grows there, its tiny leaves like miniature spatulas. In the same bed is *Asperula lilaciflora*.

The glasshouses by Beaumont are largely unremarkable except perhaps for the small fern house, dripping with moisture and with a comfortable dense verdure within. There a chunky staghorn fern, *Platycerium bifurcatum*, which gets its name from its huge antler-like fronds, is raised somewhat ignominiously on an old metal frame above a small New Zealand tree fern, *Cyathea medullaris*.

The temperature in the south-facing succulent house is breath-takingly high while the most attractive feature in the tiny Southern Hemisphere house is the central cast-iron grid which forms the floor.

Of the University Botanic Gardens in the British Isles that at Leicester is one of the most diverse and interesting, but the sad fact is that it is grossly underused by the local population.

Liverpool University Botanic Garden

Ness · Neston · South Wirral · L64 4AY
Tel. 051-336 2135

> *Open every day except*
> *Christmas Day,*
> *9 am–sunset.*
> *No entrance fee.*

In 1904 Arthur Kilpin Bulley, a wealthy, Liverpool cotton-broker, naturalist and gardener, changed the face of British gardening, when he financed a plant-hunting expedition to Yunnan in China. The expedition was led by the young George Forrest who had been recommended to Bulley by Isaac Bayley Balfour of the Royal Botanic Gardens in Edinburgh. Bulley had the time, money and enthusiasm to indulge his hobby and passion for exotic plants. The Forrest expedition was not the only one he backed. In 1911 he was responsible for sending F. Kingdom Ward to Szechwan and later Roland Cooper (who was appointed as Bulley's personal collector in 1913) to Sikkim. Bulley was also a shareholder in many other expeditions until well into the century.

His own garden at Mickwell Brow was on the Wirral peninsula in Cheshire, washed on one side by the Mersey and on the other by the River Dee. The garden, begun many years earlier, had already attracted considerable attention. Being something of a philanthropist and an active Fabian, Bulley opened his garden regularly to the public so the foreign plants he grew so successfully could reach as wide an audience as possible.

So successful were the expeditions and so obvious was the interest and enjoyment among his visitors that Bulley soon developed a nursery in the grounds. The nursery led to the establishment of the commercial firm of seedsmen, Bees Ltd., funded expressly to supply as cheaply as possible to working people many of the plants seen and admired in the garden at Mickwell Brow.

Bulley died in 1942 and in 1948 his daughter presented the garden and endowment to the University of Liverpool for a botanic garden. The only proviso was that at least part of the garden should remain accessible to the public. The garden has of necessity changed over the years and many new features have been added, notably the extensive and fine

Thousands of plants present a spectacle of colour in the fine heather garden.

Arthur Kilpin Bulley, who changed the face of British gardening.

heather garden. Bulley's inspiration, however, can still be found in the rock garden and in the extensive shelter belts created from planting holm oak, *Quercus ilex*, and Scots pine, *Pinus sylvestris*. In more recent years the ubiquitous and somehow unyielding Leyland cypress has been introduced but these, by judicious placement, in no way detract from the overall character of the garden.

The undulating topography of the site has been used to create long views, sometimes, as in the rhododendron border, 200 yards (182 metres) in length. In the heather garden the slopes have demanded more radical treatment; thousands of plants present a spectacle of colour which seems to flood across the slope.

In many respects the garden has fallen victim to its popularity and amenity value, and the priority seems to be to provide as many different features as possible, with scant regard to creating an overall cohesive effect. In the herbaceous area, for example, the long island borders enjoy neither yew or brick wall to act as a monochrome foil to the riot of colour.

The most spectacular feature at Ness is the long rhododendron walk underplanted with lilies and hydrangeas to extend the period of interest. In mild winters this begins as early as February when the magenta blooms of *R. mucronulatum acuminatum* open, and carries on well into summer. Rising behind the rhododendrons is the shelter belt of Scots

Shrubs and trees surrounding Liverpool's small stone bridge.

pine. There among the woodland flora are low peat terraces where *Pleione bulboco-dioides* and dwarf rhododendrons grow.

The garden is planted to provide visual interest all through the year. In the area devoted to specimen shrubs, to the north of the rhododendron walk, are *Corylopsis spicata* with its dangling racemes of yellow flowers in early spring, and many magnolias. Most notable is *M. thompsoniana* graced with a succession of pale cream flowers throughout summer. Maples include *Acer palmatum* 'Senkaki' and *A. griseum*, with its papery peeling bark and fiery autumn foliage. The aristocratic and yet quite common *Prunus serrula*, with its polished bark, is also there as is a thicket of *Pieris formosa forrestii*.

The terraces south of the house meet the rock garden which in late winter and early spring is a mass of flowering bulbs, including *Iris reticulata* and squills. *Phlox subulata* carpets the rocks with drifts of soft mauve and *Primula flaccida* sends up spikes of pale-blue flowers. In August *Gentiana septemfida* displays its dazzling blue flowers, followed closely by colchicums and *Cyclamen hederifolium*.

Below the rock garden are a water feature and native plant collection, where conservation is a key feature, and plants such as *Saxifraga cespitosa* and *Primula scotica* thrive.

Logan Botanic Garden

Port Logan · Stranraer · Scotland · DG9 9ND
Tel. 077-686 231

Open 1 April–30 Sept,
10 am–5 pm.
No entrance fee.

Professor Douglas Henderson, director of the Royal Botanic Garden, Edinburgh, has found his version of paradise. It is in the Logan Botanic Garden, half way along the Mull of Galloway, the southernmost part of Scotland. There, says Henderson, on a warm June evening after the public have gone, is paradise. The Logan Botanic Garden is relatively small, thirty-four acres (fourteen hectares) in extent and became an outstation of the Royal Botanic Garden, Edinburgh, in 1964. Previously there had been a private garden on the site.

The Mull of Galloway peninsula is washed on three sides by the Gulf Stream. The climate is therefore very mild and plants from the Southern Hemisphere that would otherwise not survive out of doors thrive there. The garden has an astonishing sub-tropical feel, with exotic plants such as Chusan palm, *Trachycarpus fortunei* and Australian tree fern, *Dicksonia antarctica* growing to maturity. The famous avenue of cabbage palms, *Cordyline australis*, planted in 1913, stood in the walled garden until 1979, when they were severely damaged by a cold winter. The trees have since been removed, and young plants of the species now grow in these borders. A short avenue of Chusan palms stands near the entrance.

The walled garden is the main feature at Logan. There, around a formal rectangular pool, cabbage palms grow, flowering in July with heads of fragrant creamy flowers. So prolific is *Cordyline australis*, that seedlings are being continually weeded from the borders. Equally spectacular are the Australian tree ferns, *Dicksonia antarctica*, grown outside in the British Isles in only the mildest of districts. The first of these tree ferns were planted outside here in 1912. Other Southern Hemisphere plants growing there include *Crinodendron hookerianum*, which carries the buds of its pendulous flowers throughout the winter. In spring they open into brilliant crimson lantern-like blooms.

Logan Botanic Garden has an astonishing sub-tropical feel, with many exotic plants growing to maturity.

The ancient laurel (*Laurus nobilis*) thrives there, as do drifts of *Diascia rigescens*, a recent introduction from South Africa with drifts of delightful low growing pink flowers. Close to the water's edge are swags of blue from *Geranium himalayense* 'Gravetye' and Himalayan blue poppy, *Meconopsis grandis*. Later in the year a dazzling display of African lilies, *Agapanthus*, repeat the blue theme. In the pond golden orfe move lazily, like a constantly moving sheet of beaten gold. Seeding itself freely in this area is the supremely vulgar *Echium pininiana*. This monocarpic plant from the Canary Isles flowers after three years with incredibly high spikes studded with hundreds of lavender-blue flowers.

Just north of the formal walled garden a new area has been created where the emphasis is on island beds. To the east are the rock gulley and peat garden. The peat terraces were constructed by Kenneth and Douglas McDouall, the original owners, who helped create the garden's unique Mediterranean feel. On these low terraces originated the now widely popular concept of peat gardening. Here grow primulas, lilies, *Nomocharis* and *Notholirion* and masses of dwarf rhododendron.

South of the walled garden is a bog and woodland where magnolias shelter among a planting of oak, sycamore, beech and Swedish whitebeam. The lush and succulent evergreen *Griselinia littoralis* and *Pittosporum tenuifolium* thrive.

Clianthus puniceus kakablak *from New Zealand, flowering at Logan.*

Oxford University Botanic Garden

Rose Lane · Oxford · OX1 4AX
Tel. 0865-242737

Open every day except
Christmas Day and Good Friday:
Mar–Oct Mon–Sat
8.30 am–5 pm,
Sun 10 am–12 pm, 2–6 pm;
Oct–Mar Mon–Sat
9 am–4.30 pm,
Sun 10 am–12 pm,
2–4.30 pm.
Greenhouses open daily 2–4 pm.
No entrance fee.

In winter the Oxford University Botanic Garden presents a face of uncompromising formality. The gravel paths are clearly defined and the trim lawns have a rigid symmetry. The long border beds set in the grass plots are naked and barren, taking on the appearance of freshly dug graves. There is a consistent beauty, however, and a feeling that within this walled enclosure one has entered a haven.

Bordered by the High Street to the north and by the River Cherwell to the east, there is little room for expansion or transformation. The old three-acre (1.2-hectare) walled garden seems to slip unchanged through the centuries, and indeed it is now the oldest surviving botanic garden in Britain, having been founded in the early 17th century for the advancement of medical science.

The garden owes its existence to Henry, Lord Danvers, Earl of Danby who, 'being minded to become a benefactor to the University, determined to begin and finish a place whereby learning, especially the faculty of medicine, might be improved'.[1] He bought from Humphrey Ellis, a tenant, five acres (two hectares) of damp meadow which was regularly inundated by the Cherwell. The opening ceremony took place on July 25, 1621.

The plot had been a medieval Jewish burial ground and the first priority was to raise the level beyond the reach of the Cherwell. Over the next four years, '4000 load of mucke and dunge laide by H. Windiat ye Universitie scaverger' helped keep the Cherwell at bay.[2] A tall wall was immediately erected so that the physic garden, as it was then called, should be among 'the fairest buildings of that kind in Oxford both for truth and beauty'.[3]

The magnificent gateway, designed by Inigo Jones and built by Neklaus Stone, was finished in 1632, the year before the completion of the wall. The intention that John Tradescant, the King's gardener, should be the garden's first curator never came to fruition. Instead in 1642 Jacob Bobart, keeper of the adjacent Greyhound Inn, was employed. This eccentric was a competent if unadventurous gardener. A pen portrait of him appears in an account of the garden written in 1683 by Thomas Baskerville: 'He was by birth a German born in Brunswick, As to the fabrick of body he was by nature very

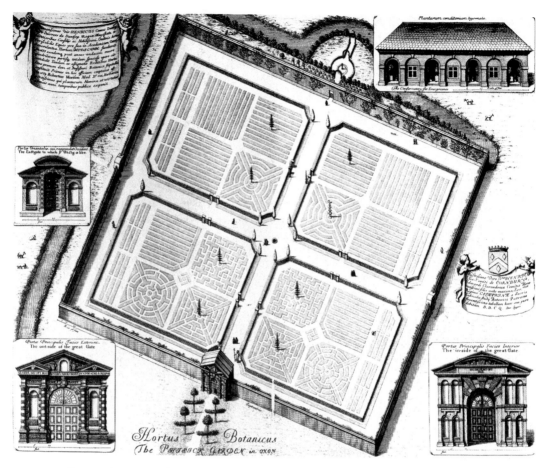

Late 17th century plan of Oxford's garden, a formal quartered square of geometric precision.

well built, tall straite and strong with square shoulders and a head well set upon them. In his latter dayes he delighted to weare a long Beard and once against Whitsontide had a fancy to tagg it with silver, which drew much Company in the Physick Garden'.[4]

Then the garden was devoted to herbs with, 'divers simples for the advancement of the faculty of medicine'.[5] Even so in July 1654 it attracted the attention of the arboriculturist John Evelyn, who wrote in his diary, '. . . went to the Physick Garden (at Oxford), where the sensitive plant was shew'd us for a great wonder. There grew canes, olive trees, rhubarb, but no extraordinary curiosities, besides very good fruit . . .'[6]

While Bobart was curator, Robert Morrison, the Scots physician to Charles II, was made Professor of Botany. He was responsible for the introduction of variegated plants which remain a feature of the garden today. A view of the garden published in 1677 affords a clear idea of what it was like. It was a formal, quartered square with each

View of the garden in 1773. Even then it was considered to have great amenity value.

quarter further divided into quarters laid out with geometric precision. There were few trees other than an avenue of clipped yews. Of these only one survives, now a huge specimen close to the south wall.

According to Baskerville, this early design was the work of Bobart, who 'first gave life and beauty to this famous place, who by his care and industry replenished the walls with great variety of trees and plants and exotick flowers, dayly augmented by the Botanists, who bring them hither from ye remote quarters of ye world'.[7]

After the deaths of Bobart in 1679 and Morrison in 1683, Bobart's son, another Jacob, took on both posts. He had none of the robust qualities of his father; he was, according to Baskerville, 'but a shrimp', in comparison. Zacharias Conrad von Uffenbach, a travelling Dutchman, who visited the garden in 1710, wrote: 'Bobart had an ugly type of countenance and an evil appearance. His nose was unusually long and pointed, eyes small and deeply sunk, mouth awry, with next to no upper lip, a great deep scar furrowed his cheek, and his face and hands were as black and coarse as those of the variest labour.'[8]

Baskerville, however, was rather more complimentary, describing him as a 'skilful and Ingenious Gardener'. One of Bobart's questionable achievements was to introduce into the garden in 1690, *Senecio squalidus*, or Oxford ragwort, from Sicily. The plant, which thrives on dry slopes, had soon naturalized itself on the tall walls of the garden and was recorded growing on those walls in 1794. Its seeds, transported by tiny parachutes, were soon taking root all over Oxford. Later, its steady march across the country was facilitated by the introduction of the railway, whose dry and well drained embankments proved ideal sites. Today the plant colonises the hard shoulders of countless motorways and is one of the first to grow on derelict building sites.

Under the younger Bobart the garden still retained its principal function of growing medicinal plants, supplemented by fruit. Although Uffenbach believed the garden 'not equal to that of Leydon or Amsterdam', even in those early days it was considered one with great amenity value. Baskerville identified the garden's dual scientific and amenity role 'prooving serviceable not only to all Physitians, Apothecaryes, and who are more immediately concerned in the practise of Physick, but to persons of all qualities serving to help ye diseased and for ye delight and pleasure of those of perfect health, containing therein 3,000 seuerall sorts of plants for ye honor of our nation and Universitie'.[9]

A few years before the younger Bobart's death in 1719, one Dr John Ayliffe took a less critical and more philosophical tour round the garden. He pronounced it 'serving not only for ornament and Delight and the pleasant Walking and Diversions of Academical Students and of all Strangers and Travellers; but of great use also . . . for the pleasant Contemplation and Experience of Vegetative Philosophy, for which is here supposed to be as good convenience as any Place of Europe (if not the best) and also for the service of all Medicinal Practitioners, supplying the Physicians, Apothecaries and who else shall have occasion for things of that nature with what is right and true, fresh and good for the service of Health and Life'.[10]

The financial well-being of the garden was secured when, shortly after Bobart's death, Dr William Sherard bequeathed to the University a herbarium and library along with £3,000. The only condition was that his friend, Dr Johann Dill, or Dillenius, should be installed as the first Sherardian Professor. This was eventually complied with in 1734.

By the end of the century the garden had become rather dilapidated. A visitor in 1824 asserted that it was once again liable to flooding and that 'the water frequently stands knee-deep above the plants', adding: 'The Oxford Garden is inadequate to the purposes of botanical instruction in the present state of science'.[11]

However, new life was breathed into it by Dr Charles Daubeny, a young man with a wide-ranging botanical interest who was made Professor of Botany in 1834. He believed that the role of the garden should be much broader than simply being a place in which to grow medicinal plants; it should be a place where plants could be studied to evaluate their scientific and economic properties. It was Daubeny who was eventually to change its title from 'physic' to 'botanic' garden.

During his first year Daubeny erected a new stove house, built two round tanks for aquatics and extended the library. He also grew the Victoria water lily and brought it into flower, only the second time the plant had actually flowered in England. His imposition of a one shilling viewing charge was less successful, causing such controversy that it was quickly withdrawn. Receptions and social functions were also held within the garden and Daubeny even resorted to keeping monkeys in a cage attached to the great Danby gateway. He was also something of a Darwinian and arranged the herbaceous plants in an evolutionary order.

In 1873 attempts were made to transfer the garden to the University Parks but the move was vetoed and the garden retained on its original historic site. As recently as 1944 was more land acquired: three acres (1.2 hectares) were leased from Christ Church and the garden extended beyond its walls into a triangle of land bordered by the Cherwell and the meadows. Funded exclusively by the University, the garden is now able to open its doors to the public every day without charge. Its main aims remain unchanged: to aid teaching and research within the University. Plants from all corners of the world are gathered; the trees have grown tall and mature. The whole garden is dominated by the tower of Magdalen College, and imbued with a lyrical spirituality, somehow removed from time. There all is safely contrived and ordered. One can walk and contemplate the plants with serenity, secure in the knowledge that the wilderness is excluded, and experience the 'subtle psychological nexus between the Garden Spirit and the soul of Universities and Academies – the classic and sacred Groves of Thought, Learning'.[12]

The living textbook approach adopted by many of the old botanic gardens, with plants displayed in regular, geometric beds, is still very much in evidence at Oxford. There, within a neatly ordered square of land where the lawns are dissected by gravel paths, can be found a lush verdure of regimented planting. Just occasionally, in the south-west and south-east corners, the planting is freed from rigid formality.

In these beds dense clumps of *Geranium versicolor*, *Geranium procurrens* and a number of hostas crowd the ground merging into one another. Over their heads towers a curious and ancient Kentucky coffee tree, *Gymnocladus dioica*, the seeds of which it is said were roasted to make a coffee-like drink in its native America. The tree has a gaunt ragged appearance until well into spring. *Magnolia soulangeana* is there, its blooms like plump pink birds carried until late spring, and at its feet grow *Tellima grandiflora* and cuckoo pint, *Arum maculatum*, with its cowled flowers. There are drifts of tall growing *Geranium psilostemon* and the vigorous spreading *Geranium macrorrhizum* with its beautiful pink petals and crimson calyces.

The irregularity of this corner is further enforced by the clumps of bamboo that grow nearby, some of which struggle along at knee height while others, such as *Arundinaria nitida*, grow into a superbly elegant form. This gloomy thicket of bamboo merges into a

Oxford, the oldest surviving botanic garden in Britain, showing (right) the Danby Gate which was designed by Inigo Jones.

miniature meadow where the grass has been allowed to grow long. Sweet cicely, *Myrrhis odorata*, lifts its elder-scented white flowers high above grass studded by buttercups and daisies. Nearby is a curious low-growing shrub, *Dipelta floribunda*, its scented flowers not unlike those of foxglove and its bark peeling in crinkly brittle flakes.

So soundly constructed were the original walls around the garden that they stand strong to this day. Occasionally *Senecio squalidus*, the notorious Oxford ragwort, can be seen high among the upper masonry. More attractive plants haunt the wall on the western boundary: *Solanum crispum*; *Syringa* x *persica* 'Alba', draped in exquisite white flowers; *Actinidia kolomikta*, its leaves splashed with delicate pink and white; *Akebia trifoliata*, the sinuous twining climber from China; *Ceanothus dentatus*, straining against the grey stone; and *Xanthoceras sorbifolium*, another native of China.

These climbers command the garden. At their feet a long, almost too narrow, herbaceous border runs the length of the garden, rich in *Chrysanthemum maximum*, kniphofia, nepeta, coreopsis, penstemon and blue geraniums in broad drifts.

Across from the gravel paths are the lawns with their richly planted order beds, including beds of euphorbia and beds of herbs. Among the latter are mint and feathery fennel, angelica, and spices and flavourings from around the world. There is also a clump of delicious lovage. John Parkinson in *Theatrum Botanicum*, published in 1640, describes with customary wonder this plant and its leaves 'of a sad green colour, smooth and shining from among which rise up sundry strong and tall hollow green stalkes five or six foote high, yea eight foot high in my Garden'. Parkinson does not mention the plant's culinary application, though Gerard, in the improved 1636 *Herball*, describes, 'the people of Gennes in times past did use it in their meates, as wee doe pepper'.

In a nearby bed are deadly nightshade, wormwood, datura and aconite. Nearby is *Arundo donax*, a strident giant reed that reaches to a height of ten feet (three metres) or more.

Bulbs have their place under the amazing *Davidia involucrata*. To see this tree flowering in spring is to share the fascination that the early plant hunters must have felt when they first saw it flowering in its native China. The flowers, tight small crimson whorls, are insignificant. What makes the tree so magical are the white bracts that surround the flowers, earning the tree its common name, handkerchief, or dove, tree.

Near the fountain which marks the axis of the cruciform and which is dwarfed by mighty trees is a bed of *Petasites albus*, a plant thought far too vigorous by most gardeners. It is a characteristic of Oxford, however, that even the most vigorous plant has a place. *Polygonum sachalinense* and *Polygonum cuspidatum* are at home as much as the horsetail ferns which relentlessly attempt to spread. Above the south-east corner rises a most stately *Pinus nigra*, its foliage pressed against the sky.

The bed which flanks the east boundary is devoted to ferns: hart's tongue, producing its erect fronds early in the year; shield fern, as soft as cat fur; and *Dryopteris affinis*, nut brown and sage green. There are also fresh green shuttlecock ferns which rise from thick woody crowns. Nearby are beds devoted to geraniums, including *G. versicolor* with reflexed petals, and *G. pratense*, the distinctive native meadow cranesbill.

On the other side of the eastern wall are the glasshouses full of exotics. Grapefruit, palm, ginger and lemon grow in the largest house, and from the roof hang iridescent curtains of bougainvillea. Bananas fruit in the water-lily house and sugar cane grows with its roots in the water. In the succulent house are bird-of-paradise, *Strelitzia reginae*, an olive tree and a date palm. There is a bench devoted to carnivorous plants, displayed so that close-up views can be had of the minute jewel-like drops on the stems of the sundew, *Drosera*. In no other botanic garden have I found such plants so accessible.

To the south the garden has expanded onto a triangle of ground bounded by Christ Church meadow and the Cherwell. In this area can be found a deep herbaceous border, a perfectly proportioned rock garden and a rose collection demonstrating the origin of garden roses. One delight is to see the old *Rosa damascena versicolor*, the striped York and Lancaster rose, which looks like a crushed silk handkerchief.

What is so alluring about this garden is its compact size, the staggering diversity of the plant collection, and the peace and tranquillity contained within its high walls.

Nuneham Courtney

Nuneham Courtney · Oxon
Tel. 0865-242737
Open weekdays · Apr-Sept · 9 am – 5 pm · No entrance fee*

In recent years the University has established a fifty-acre (twenty hectares) arboretum at Nuneham Courtney, expanding an earlier eight-acre (3.2 hectare) pinetum begun by Edward Harcourt, Archbishop of York, between 1830 and 1844. There rhododendrons, azaleas and camellias flourish beneath a mixed planting of deciduous trees and conifers. There is a bluebell wood, a heather garden and a fine collection of oaks, linked together by serpentine walks. The soil also supports a rich diversity of woodland flora and trilliums appear to flourish.

1. R. T. Gunther, *Oxford Gardens*, Oxford, Parker and Son, 1912.
2. Ibid.
3. Ibid.
4. Ibid.
5. Ibid.
6. Sir William Temple, *Gardens of Epicurus*, Chatto and Windus, London, 1908.
7. Ibid.
8. As quoted in R. T. Gunther's *Oxford Gardens*.
9. Ibid.
10. Ibid.
11. Ibid.
12. Introduction by A. F. Sieveking to Sir William Temple's *Gardens of Epicurus*, Chatto and Windus, London, 1908.

The Royal Botanic Garden, Edinburgh

Inverleith Row · Edinburgh · Scotland · EH3 5LR
Tel. 021-552 7171

Open every day except New Year's Day:
Mon–Sat 9 am–sunset,
Sun 11 am–sunset.
No entrance fee.

In 1670 Robert Sibbald, the first Professor of Medicine at Edinburgh University and his friend, Dr Andrew Balfour, established a physic garden on a small plot of land attached to St Anne's Yard, Holyrood Abbey. Their encounters with quack medical practitioners in contemporary Edinburgh and with inferior drugs based on fanciful concoctions rather than on sound scientific principles had previously led both men to grow their own medicinal herbs. In the small physic garden at St Anne's Yard their own collections were united with that of one of Sibbald's wealthy friends, Patrick Murray, to establish what is now the second oldest botanic garden in Great Britain. (The oldest is at Oxford, see page 95.)

The physic garden was a mere forty feet (twelve metres) square and soon became far too small to accommodate the 800 or so plants growing there. In 1672 a larger area of ground attached to Trinity Hospital about half a mile (.8 kilometre) from Holyrood was therefore acquired. This relatively large rectangular garden was nurtured by head gardener James Sutherland, and contained specific areas for flowers, aquatics, medicinal plants and trees. The garden flourished and by 1683 Sutherland was able to publish a catalogue of its 2,000 plants. In it he highlighted the didactic role of the garden while simultaneously hinting at the poor state of medicine in Edinburgh. He wrote, 'Apothecaries Apprentices could never be competently instructed in the Knowledge of Simples (which necessarily they ought to be) before Establishing of this Garden; for now they may learn more in one Summer, than formerly it was possible for them to do in an Age'.

The garden thrived along with the city and in 1695 it expanded to take in part of the Royal Gardens at Holyrood. Despite various vicissitudes it was another sixty-eight years before cramped conditions and city pollution forced John Hope, the current Regius Keeper, to go in search of another site where he could unite both gardens. In 1763, having acquired a five-acre (two-hectare) site to the north at Leith Walk, the entire plant collection was moved.

The tenure at Leith Walk lasted for fifty-seven years, but by 1820 it, too, was considered too small for the vast collection of plants established by William McNab, the

Principal Gardener, whose collecting enthusiasm since 1810 had known no bounds. Again the entire plant collection was moved, this time over a period of three years. Mature trees were dug up, the largest being a forty foot (twelve metre) high weeping birch, and trundled through the streets on McNab's unique transplanting machine. Their destination was to be the new fourteen and a half acre (5.8 hectare) site to the north at Inverleith, the current home of the Royal Botanic Gardens.

Throughout the nineteenth century more land was gradually acquired. In 1874 John Hutton Balfour, the then Regius Keeper, felt he, too, did not have sufficient space for an arboretum. When an adjacent twenty-eight acres (eleven hectares) of land became available around Inverleith House he successfully lobbied for their purchase. In 1876 the Botanic Gardens embraced this additional land, and the garden began to assume the layout it has today.

The actual integration of the ground fell to Balfour's son, Isaac Bayley Balfour, University Professor of Botany and Keeper of the Garden, in 1888. Under his direction the garden really began to take shape and he, perhaps more than any other Keeper in the garden's history, along with the plant hunter, George Forrest, should be credited with giving the garden the character we perceive today.

The reasons are simple enough. As the nineteenth century drew to a close, plant hunters began to explore the remote regions of China, Nepal and Tibet. Some, such as Augustine Henry, fell upon the activity by accident. Although qualified as a doctor, Henry found the work uncongenial. He applied for and secured a post with the Chinese Maritime Customs at Ichang on the Yangtze River. There he was overcome by the wealth and variety of the local flora but frustrated because his scant botanical knowledge meant there was little he could identify. Determined to remedy this, he sent back to Kew in 1886 a collection of dried specimens. Kew responded with excitement. They encouraged Henry to collect more and, by the time he left China in 1900, he had sent back to the herbarium at Kew more than 158,000 dried specimens.

Astute nurserymen of the day were quick to see the commercial possibilities of introducing to British gardens the living counterparts of the dried specimens. Men such as Sir John Henry Veitch were greatly impressed by Henry's collection at Kew and became determined to introduce anything that took their eye. In March 1899, Veitch mounted an expedition to China, headed by the now legendary Ernest Henry Wilson, to search out the seeds of the spectacularly beautiful handkerchief tree, *Davidia involucrata*, originally found and named in 1869 there by the French missionary, Armand David.

The garden at Edinburgh has the largest and most comprehensive rock garden concentrating on species in Britain.

Once in China Wilson made contact with Henry, who was able to tell him where the handkerchief tree was, although it had been twelve years since he himself had seen it. When Wilson eventually found the tree he discovered to his horror that it had been cut down and all that remained was a stump. However, more living specimens were soon found nearby and the seeds sent back.

All this botanical excitement was going on while Balfour was consolidating the garden at Edinburgh. His frustration at not being able to augment his own herbarium must therefore have been considerable. In 1904, the first of the great twentieth-century plant collectors, Arthur Kilpin Bulley, Liverpool cotton-broker, naturalist, enthusiastic gardener and founder of the seed suppliers, Bees Ltd, approached his friend Balfour for advice as to who to send to Western China on a plant-hunting expedition. Balfour had no hesitation in recommending the young George Forrest, undoubtedly seeing the expedition as an opportunity for Edinburgh to acquire its own Sino-Himalayan collection while assisting in the introduction to British gardens of many hitherto unknown species.

Forrest, who had earlier been on unsuccessful adventures to Australia and South Africa in search of his fortune, was then working in a menial post in Edinburgh's herbarium. After two years there he had at least gained a working knowledge of plants from around the world.

The expedition proved to be fabulously successful and, on his return in 1907, Forrest brought with him thousands of herbarium specimens and many seeds which Balfour and his staff helped clean and distribute to the expedition's backers. Balfour knew that the rhododendrons, meconopsis, camellias, and primulas that Forrest introduced would change the face of gardening at Edinburgh and in British gardens generally.

Forrest returned to China several times. By the time of his death in 1934 he had sent back to Edinburgh more than 30,000 herbarium specimens and many hundreds of seeds, besides supplying rhododendrons direct to enthusiasts such as John Charles Williams. The latter's collection at Caerhays Castle in Cornwall included more than 250 of Forrest's introductions.

Ironically, climatic conditions at Edinburgh were not ideal for the cultivation of rhododendrons and Balfour always hoped to acquire a garden in the west of Scotland where the plants could be grown in an almost natural environment. His dream, however, never materialized. He died in 1922 and it was left to his successor, William Wright Smith, who shared a similar passion for plants from the Sino-Himalayan region, to establish the Younger Botanic Garden in the west of Scotland. Even so, many of Forrest's introductions do thrive in Edinburgh, and the latest count of *Rhododendron* species is in excess of 400, although not all of them were introduced by Forrest.

It is this wealth of rhododendrons and their spectacular display that makes the Royal Botanic Gardens the third busiest tourist attraction in Scotland. In 1984 it attracted more than 700,000 visitors; the record number through the gates in any one day topped 17,000 in June 1985.

Although the garden covers only seventy acres (twenty-eight hectares), its gently undulating topography, and its vast number of trees and shrubs creating natural screens which restrict sweeping vistas, give the impression of a much larger and more rural site. One is hardly aware of the proximity of the city which encroaches on all sides; Edinburgh only becomes evident when viewed from the higher ground at the centre of the garden. This natural topography has the benefit of giving the garden a more satisfying cohesive unity than the Royal Botanic Gardens at Kew.

If the garden boasts more than 400 species of rhododendrons, it also prides itself on having the largest and most comprehensive rock garden concentrating on species in Britain. This is another legacy of collecting and of the complete rebuilding programme carried out by William Wright Smith and Isaac Bayley Balfour between 1908–14. It was then that the rock garden assumed its present shape although a water course was added later. Conceived in as natural a style as possible, the design of the rock garden was a deliberate attempt to evoke the escarpments of many of the plants' Himalayan origins, as described in Forrest's diaries. The peat garden was developed in 1939 to take the many plants that were unhappy in the rock garden because of their need for cooler and moister conditions.

The rock garden is not without its critics; its over-contrived neatness, artificiality and huge scale can sometimes detract from the plants. It does, however, display ingenuity in planting, with shrubs such as *Cytisus ardoinii*, genista, cotoneaster, potentilla and several *Juniperus* species helping to conceal the rocks. Many common plants are used, and many of the juxtapositions, such as *Cotoneaster horizontalis* with its brilliant red autumn berries set beside a glaucous, *Picea pungens* 'Glauca pendula' may have become too well known. Yet they are no less successful for that, and well worth repeating. These shrubs, together with many other slow-growing conifers and dwarf rhododendrons do provide a variety of leaf form and structure which extends the visual interest throughout the year.

Paths stitch their way between rocky outcrops and around tiny crevices where lichen clings to vertiginous surfaces and the sound of tumbling water lays a delightful trace over the scene. From late February the garden bursts into colour from bulbs, alpines and *Rhododendron forrestii*, with its bell-shaped crimson blooms. In early summer *Primula florindae* grows tall while in autumn the silver-blue domes of *Artemisia schmidtiana* 'Nana' are crowned by tiny yellow flowers.

Below the rock garden the water course opens out into a more substantial stream before emptying further down the slope into a small pond. This stream is a recent addition to the garden. Begun in late 1984 it has become a fine example of what can be achieved in so short a time with the banks already dense and lush with astilbe, hosta, dicentra and rhododendron. At the waterside grows *Gunnera manicata* and *Cornus alba* with its vivid burgundy stems. An ancient, seemingly neglected punt floats on the still water, ducks swim and reflections sparkle.

The stream at Edinburgh, its banks already dense and lush with plants.

The lushness of this area is repeated in the woodland garden where herbaceous plants shelter beneath eucryphias, camellias and a mass of rhododendron. One eye-catching example of ground cover here is *Maianthemum kamtschaticum*, the white flowers of which cover the woodland floor in spring. *Anemone nemorosa* is here, too, as is the greater celandine in profusion.

Beyond the woodland garden, the ground falls gently away through the peat garden before climbing steadily up through the arboretum and on to Inverleith House. The peat garden drops in a series of irregular terraces supported by peat turves. Many dwarf rhododendrons grow here as well as schizocodons, the flowers of which have charming, deeply fringed edges. The discreet adder's tongue fern, *Ophioglossum vulgatum*, flourishes, its curious undivided frond cowled around the snake-like tongue which gives the plant its name. The astonishingly beautiful orchid, *Dactylorrhiza elata*, from south-west France and North Africa, thrives, with the pure white, waxen and fragile *Trillium grandiflorum*.

The garden's definitive collection of rhododendrons grows along the walk that runs from the West Gate in a long arc around Inverleith House. The borders on each side of the walk are packed with rhododendrons, providing shelter for spectacular displays of hostas, meconopsis, lilies and primulas. There are also some fine clumps of bamboo close by and from there views can be had across the oak copse to the long herbaceous border.

This immensely long and deep border is backed by a beech hedge of equally impressive proportions. In summer the view along the border is one of the highlights of a visit to the gardens. The construction of such colourful borders is an art currently being practised throughout Britain's botanic gardens and one which teaches plant association in the best possible way, by example.

On the other side of the beech hedge, through an archway and beyond a border of brightly coloured annuals, is the demonstration garden, established in 1962 and protected by holly hedges. Here such principles as pollination and relationships among plant families are set out. It is a fascinating kaleidoscope of nature's ingenuity. Poisonous plants have their place: thorn-apple, *Datura stramonium*, and deadly nightshade, *Atropa belladonna*, the black shiny seeds of which can be fatal. The lethal potential of plants is also present in some which, used properly, can have a beneficial medicinal effect. Henbane, *Hyoscyamus niger*, for example, was used by Crippen to murder his wife in 1910. Other medicinal plants represented include *Nicandra violaceae*, with its black stems and curious five-winged calyces clasping the fruit within. The herb bed basking in the sun provides the scents of the Mediterranean: marjoram, chervil and thyme.

Beyond the herbaceous border the roof of the temperate palm house rises above the canopy of mature trees. This house, opened in 1858, is the least aesthetically satisfying building in the gardens. With its ruthless emphasis on the vertical, the impression gained is of tall columns rising between immense windows on top of which sits an entablature and dauntingly high-stepped curvilinear glass dome. The immense, awkward structure creates a mausoleum-like effect which fails to derive any real eloquence from the language of Victorian conservatory construction which it uses liberally.

Several specimens of the palm *Livistona australis* grow with some vigour – the largest currently pushing against the roof glass – while the camphor tree, *Cinnamomum camphora*, is of more modest proportions. The oil of this plant is said to act as an insect repellent; Marco Polo records how, in thirteenth century China, moth-proof cabinets were made from its wood.

Connected to the temperate house is the octagonal tropical palm house, a place of clammy tropical heat. One of the oldest plants in the garden, the thatch palm, *Sabal blackburneana*, grows there; it was brought from the Leith Walk Garden when the contents were transferred to the present site in 1822.

Behind the tall Victorian palm houses are the low, neat tropical peat and rock houses, which have been cleverly landscaped internally to give the visitor an idea of the plants' natural environment. The varying levels inside the buildings have allowed the construction of high peat walls and rocky chasms. Epiphytic orchids, such as *Dendrobium* and *Brassia*, grow on artfully arranged tree trunks, and produce flowers like large exotic winged insects. Vines hang from internal wires; the air potato or climbing yam, *Dioscora bulbifera*, is a strange looking plant. Dutchman's pipe, *Aristolochia*, is also there, seducing insects into its flowery bowls and only releasing them when those bowls wither.

In the rock house, the atmosphere is hot and sticky with lush vegetation springing from the ground. There is a large collection of ginger, *Hedychium*, and a rocky gorge festooned with African violets. There is also a sealed glass case with insectivorous plants. The immature pitchers of *Nepenthes* x *Ratcliffeana*, are said to be hermetically sealed by a dense growth of interweaving hairs: the fluid inside is thus sterile, and is used in Borneo to bathe sore eyes.

One of the world's latest ecological discoveries can also be found here. The jojoba, *Simmondsia chinensis*, produces oil which can withstand high temperatures and is therefore used in a number of products, such as soap and shampoo, as a substitute for sperm whale oil.

The most exciting recent development at Edinburgh has been the opening, in 1967, of the exhibition plant house. Constructed to afford virtually unimpeded internal space, the glasshouse is 420 feet (130 metres) long, 60 feet (18 metres) wide and up to 36 feet (11 metres) high and has six internal landscaped compartments. Each has its own micro-climate, ranging from desert terrain to tropical rain forest.

The cactus and succulent house was contrived to demonstrate the parallel yet separate evolution of plants from arid zones. It opens onto the temperate and aquatic house, dominated by a central pool. This is crossed by a bridge hung with the diurnal flowers of the blue dawn flower, *Pharbitis learii*. After the dryness of the cactus house, the hot damp atmosphere of the temperate and aquatic house takes one's breath away. Moisture seems to run from every surface and the flowering clock vine, *Thunbergia grandi flora*, with its blue orchid-like flowers, scrambles in every direction. The glasshouse is swamped by the delicious smell of frangipani, *Plumeria rubra*. The long strap-shaped leaves and tall exotic flowers of the Brazilian *Vriesea imperialis* reach out over the pool and the wonderful arrow-like leaves on long blue-grey stems of the *Xanthosoma violaceum* make amazing accents among the dense mass of green foliage.

The exit from the warm temperate house opens into a gallery that encircles the temperate plant house and allows close examination of the foliage of the Australian trees growing there. Acacias and eucalyptus rub shoulders with *Cupressus cashmeriana* from south-west China; the latter sports breath-taking foliage hanging from tawny branches. Sheltering beneath these trees are many beautiful shrubs such as the bottlebrush, *Callistemon*, with its superb flowers. Its seeds can remain in a viable condition for many months until scorched by a forest fire. Only then are they released from the plant to germinate on the freshly burnt ground. The tree tomato, *Cyphomandra befacea*, from Peru is also there, its fruits, which can be eaten raw or made into a jelly, hanging like solid green lanterns. Over all hangs the intoxicating, almost sickly fragrance of *Acacia crassiuscula*.

The tropical aquatic house contains one of the most curious plants in the garden: the bull's-horn acacia, *Acacia spadicigera*. It is unbelievable that any tree with such vicious double thorns could need protection in the wild but apparently it does. It lives a symbiotic life with a species of aggressive ants which defend the tree from clinging plants such as vines and in return feed from its nectar. The main attraction, however, is undoubtedly the giant leaves of the wondrous Victoria water lilies, which float on the still surface of the pool. The most spectacular view comes from the chamber beneath the pond which allows close examination of the magnificent undersides of the leaves. Their incredible fluted structure, designed to spread the load of the vast leaf, is said to have inspired Paxton in his designs for the grand Victorian glass and iron conservatories that he pioneered. To stand beneath these beauties and see the light shimmering through the translucent leaves is an extraordinary experience.

The visitor overwhelmed by this sub-aquatic spectacle will be brought back to earth by a few minutes spent in the fern house. For here in this cool atmosphere, where the Australian tree fern, *Dicksonia antarctica*, grows tall and the *Cibotium glaucum* from Hawaii spread like giant bracken with fronds as huge and fat as bishop's croziers, the visitor is caught in a shower when, unannounced, the mist sprays burst into life.

Dawyck Botanic Gardens

Near Stobo · Tweeddale · Scotland
Open daily 1 April–30 Sept · 9 am–4 pm · Entrance fee

Recently a further outstation has been added to Edinburgh's portfolio of gardens. Dawyck Botanic Garden, 28 miles (45 kilometres) from Edinburgh, is an arboretum to which Edinburgh is currently adding trees and shrubs of wild origin.

The first known planting at Dawyck dates from circa 1680. For 200 years the estate was owned by the Naesmyth family whose vigorous planting policy attracted the attention of John Loudon who mentions the estate in his 1838 book, *Arboretum et Fruticatum*. In 1897 the arboretum passed into the Balfour family who planted the extensive collection of rhododendrons. In 1979 the arboretum was given to the nation to subsequently form part of the national botanical collection.

Some fifty acres (twenty hectares) in extent, Dawyck boasts a rich mixed collection of deciduous trees and conifers with thousands of daffodils and rhododendrons flowering in spring, the latter of which includes introductions by the plant collectors E. H. Wilson and George Forrest.

The Royal Botanic Gardens, Kew

Kew · Richmond · Surrey · TW9 3AB
Tel. 01-940 1171

*Open every day except
Christmas and New Year's
Day,
10 am–dusk.
Entrance fee.*

The Royal Botanic Gardens, Kew can justifiably claim to be the finest botanic garden in the world. Its origins lie in the nine-acre (3.6-hectare) garden started in 1759 by Augusta, Dowager Princess of Wales and mother of George III. Augusta lived in what was known as the White House with her husband, Frederick, eldest son of George II. The house had once been the home of Sir Henry Capel, whose garden had been described by John Evelyn in 1670 as having 'the choicest fruit of any plantation in England' with 'oranges and mirtles' in two greenhouses.[1]

Frederick was an ambitious gardener and when he died in 1751 his enthusiasm was carried on by his widow with the help of Lord Bute, who was to become Prime Minister in 1762. Bute encouraged Augusta's garden improvements and was instrumental in employing the architect Sir William Chambers to re-landscape the grounds and to erect a number of eye-catching buildings. The most notable of these remains the 163 foot (49 metre) high pagoda, built with astonishing speed between the autumn of 1761 and the spring of 1762. So rapid was its construction that Horace Walpole, who lived at Twickenham, noted in a letter to Lord Stafford, 'We begin to perceive the Tower at Kew from Montpellier Row: in a fortnight you will see it in Yorkshire'.

When it was finished the pagoda's projecting roof was covered with 'plates of varnished Iron of different colours; All the angles of these roofs are adorned with large dragons, being eighty in number, covered with a kind of thin glass of various colour.'[2]

It was the Chinese ability to evoke such moods as melancholy and gaiety that attracted Chambers who had visited Canton earlier in his life and was to become the great champion of chinoiserie. The fascination for Chinese curiosities was not new, however. A hundred years earlier Sir William Temple, diplomat, amateur philosopher and gardening writer, had written of Chinese landscapes. The same curiosity about chinoiserie was to persist in the public imagination throughout the eighteenth century while the English landscape garden developed to its fullest maturity in the hands of Capability Brown.

Chambers did not confine himself to China in his search for inspiration. He further embellished the grounds with an odd assortment of Greek, Indian, Gothic and Turkish

The Palm House in Kew Gardens, mid 19th century.

buildings, using them to create a number of focal points. So pleased was he with his achievements that in 1763 he published the plans in a book dedicated to Augusta, Princess Dowager of Wales.

Chambers was also responsible for the Orangery, a conservatory of typical eighteenth-century style, with windows to one side and opaque roof. This building has never really been suitable for plants. It is now used as an exhibition centre but is still rather unsatisfactory. In Chambers's words, 'The front extends one hundred and forty five feet: the room is one hundred and forty two feet long, thirty feet wide, and twenty five high. In the back shed are two furnaces to heat flues laid under the pavement of the Orangery, which are found very useful, and indeed very necessary in times of hard frost'.[3]

Without doubt the gardens at Kew aspired to be the finest in Europe. The physic or exotic garden with plants 'collected from every part of the globe, without regard to expence' would be 'the amplest and best . . . in Europe', Chambers continued.

Responsibility for the planting was firmly in the hands of Lord Bute and, from 1759, William Aiton, who had trained at the Apothecaries' Garden at Chelsea. In 1789, Aiton was able to list 5,535 plants under cultivation in the garden in his catalogue *Hortus Kewensis*.

When George III succeeded to the throne in 1760 he engaged Capability Brown to landscape the grounds of Richmond Lodge, his home which was adjacent to his mother's. On Augusta's death in 1772 the King united both estates and the formation of Kew as we know it today began.

The great botanist, Sir Joseph Banks, was the first, if unofficial, director of Kew and it was he who established the garden's scientific reputation. After travelling with Captain Cook to the South Seas he became a scientific entrepreneur, sending plant collectors to all parts of the world. The famous voyage of 1787, led by the ill-fated Captain Bligh aboard HMS *Bounty*, was essentially an expedition to collect breadfruit from Tahiti for cultivation in the West Indies. The voyage was interrupted by the mutiny and David Nelson, one of Kew's gardeners who had sailed with Bligh, took the literally fatal decision to side with the captain; he was cast adrift in an open boat by the mutineers and eleven weeks later was washed up on Java where he died.

By this time Kew's blossoming reputation was such that botanists and collectors had begun to send seeds and plants there in preference to the Apothecaries' Garden at Chelsea, whose standing as the centre of botanical excellence was in decline. In 1820 both George III and Sir Joseph Banks died and the horticultural direction of the garden was in the hands of William Aiton's son, William Townsend Aiton, who had succeeded his father in 1793. George IV rarely visited Kew during this period and the lack of such royal patronage brought about a slight decline in the garden's reputation. Its fortunes were transformed when, in 1840, it was handed over to the Commissioner of Woods and Forests. Sir William Hooker was appointed in 1841 with immediate responsibility for the eleven-acre (4.4-hectare) botanic garden. In 1845 he also took control of the adjacent pleasure-grounds. The landscape designer William Andrews Nesfield was employed to turn the gardens into a unified whole.

Nesfield used the palm house designed by Decimus Burton and Richard Turner and completed in 1848 as a central focus, and contrived a scheme of long rides and broad vistas to cut the site in a dramatic and bold way. In September 1848, *The Illustrated London News*, reporting the opening of the 'Great Palm House', quoted Hooker's 'admirable Guide to the Gardens'. '. . . From the great western entrance of the palm-stove three vistas will radiate at equal distances, commanding views through the pleasure-grounds; one, inclining to the south, in the direction of the Pagoda; the second, or western vista, towards the river and woods of Syon; and the third towards Brentford. From the south-eastern angle of the palm-house the walk is continued round the water; and from the opposite side the best view of the structure may be seen, and its reflection in the lake.' Syon Vista, Pagoda Vista and Broad Walk have now grown to lush maturity, creating a number of spectacular views which embrace almost the whole length and breadth of the gardens.

Under Hooker the garden enjoyed increasing public acclaim. The annual number of visitors, aided by longer opening hours, swelled from 9,000 in 1841 to 64,000 in 1848, and from 124,000 in 1849 to 500,000 in 1865.

View of Pagoda Walk, one of the spectacular vistas created by the landscape designer William Andrews Nesfield.

By the time of Hooker's death in 1865 the garden, now occupying about 250 acres (101 hectares) was well stocked and established, and the Temperate House was well under way. The latter was also designed by Burton (as are the main gates) and in its day formed the largest greenhouse in the world. The central section with two octagons was completed in 1862 and the north and south wings were added in 1895. Two years later, on the diamond jubilee of Queen Victoria, the acreage of Kew was further extended when the Queen presented the woodland around Queen's Cottage to the garden, provided the woodland remained in its semi-wild state.

Kew's notable past economic successes have included rubber and quinine. Sir Joseph Hooker had the seeds of the Para rubber plant, *Hevea brasiliensis*, brought from South America in 1876 and the seedlings from these were sent on to Ceylon and Malaya. Aided

*The formal parterre of dwarf box, centred round an old Venetian well-head, in the
Queen's Garden at Kew.*

by advice from Kew, plantations were soon established in the Far East and from these
have developed the present 10.5 million acres (4.25 million hectares) used in rubber
production today.

Kew's success with quinine can truly be said to have altered the history of the world,
conquering, as it did, malaria. The drug quinine is obtained from the bark of several
species of the South American tree, *Cinchona*. By the close of the nineteenth century
supplies had become erratic. Trees were being decimated in the pursuit of their bark and
the stock was not being replaced. Seeds raised at Kew in the 1860s helped establish the
tree in India, and alleviated the scarcity of the drug.

Kew's resources are so vast that it is better to describe individual aspects rather than attempt a general tour of the grounds, which can only be fully appreciated after many visits. The rhododendrons that flood the rhododendron dell with brilliant colour in spring are of breathtaking beauty, as are those that surround the Aeolus Mound near the pond, and the main species collection which flanks Cedar Vista. The bluebells in Queen Charlotte's Cottage's grounds are spectacular while the rock garden affords hours of delight. The crab apples and hawthorns just south of the Temperate House form a drift of sugary pinks and whites; the camellias along Kew Road are sharp and brilliant. In the azalea garden, banks of shimmering colour are set off by subtle greys while close by are masses of bamboo and an avenue of chestnut, fiery with red and white candles. The trees, especially, have an air of majesty as they droop their skirts down to the lawns, and as the conifers pierce the canopy.

Yet my favourite corner of the gardens is definitely that single acre (.4 hectare) tucked behind Kew Palace. There, in what is known as the Queen's Garden, has been created a seventeenth-century idyll. The palace, originally called the Dutch House, was built in 1631 for Samuel Fontrey, a rich London merchant. In 1959 the director of Kew, Sir George Taylor, decided to create a garden contemporaneous with the house. Early gardening books were consulted and the resulting plan contained all the elements of a seventeenth-century garden with the emphasis on small individual areas and changes of levels.

A seventeenth-century garden was described in 1676 by the writer William Lawson[4] as a place where man's senses could be refreshed and his weary spirit renewed. It was somewhere to walk and savour the fragrances of the herbs grown for culinary and medicinal uses. His concept was both philosophical and utilitarian, envisaging an environment where, within a rigid framework, separate enclosures contained all that was necessary for man's enjoyment.

Knots, topiary, 'daintie Herbes, delectable floures, pleasant Fruites, and fine Roots,' Lawson wrote, were to be found in most gardens by the close of the century. In the Queen's Garden are formal parterres, a boscage (a hedge on stilts, made up of trees clipped in such a way as to leave the lower trunk exposed) of hornbeam, clipped hedges of yew and box, a sunken nosegay garden and a mound. Within the formal, almost claustrophobic space, the garden is crammed with a rich and fascinating variety of plants, all of which would have been known to the seventeenth-century gentleman.

The dominating feature is the mound, its slopes clad with close-clipped dwarf box through which a path cuts a sinuous spiral. A wrought-iron rotunda sits arrogantly on top and it is from there that the gentleman would have surveyed his precisely ordered world, cut off from any surrounding wilderness by walls and hedges.

He would have viewed the rigidly formal parterre of dwarf box infilled with rosemary, sage and curry plant, centred on an old Venetian well-head. He would have taken in the avenue of severely pleached hornbeam, but the sunken nosegay garden would have been out of sight, tucked behind a tall yew hedge.

A raised laburnum walk flanks three sides of the nosegay garden, its flowers hanging

The new, carefully landscaped Princess of Wales Conservatory, sited near the rock garden.

down in a dazzling golden shower in May. The lower paths are edged in dwarf box, the centre beds filled with a riot of herbs and flowers, many of which feature in the early herbals – wormwood, sage, pinks and hellebores.

The features that dominate Kew, however, are the glasshouses, which are the epitome of Victorian engineering and inventiveness. Viewed from a distance, Burton and Taylor's palm house, built between 1844–8 and surrounded by formal flower beds, seems to float on the adjacent pond. Within the cathedral-like space bananas fruit, palms stretch up to the roof and exotics grow untouched by the harsh winter outside. *Encephalartos longifolius*, probably the oldest living greenhouse plant at Kew, was collected in 1775 by Francis Masson in South Africa, and now leans on a steel jack for support. A member of one of the Cycads families, this plant has long fronds and large seed cones.

Such glittering cathedrals of glass also contain the seeds of their own destruction. The different expansion rates of glass and iron meant that they were in need of constant attention. The palm house was substantially repaired in 1959 and has been almost completely dismantled and reconstructed from 1985–7.

Grander in an altogether different way is the other great glasshouse, the temperate house, in its day the largest glasshouse in the world. It post-dates the palm house by eighteen years; construction spanned more than thirty years, from 1860, when the main block and octagons were built, to 1898, when the south and north wings were completed.

The temperate house resembles a glittering wedding cake, with its external stucco piers topped by ornate urns, each crowned by a stucco cornucopia. In the southern wing grows that most exotic genus of plants, *Strelitzia*, named after Charlotte, wife of George III, of the House of Mecklenburg-Strelitz. The largest displayed is *S. nicolai*, flaring a good fifteen feet (five metres) into the air and crowned with great beak-like blooms.

In the main block can be found the largest plant growing under glass in the world, the Chilean wine palm, *Jubaea chilensis*. Grown from seed collected from Chile in 1846, its thick chunky trunk towers above the surrounding foliage. Ancient ginger jars are home for small tree ferns, *Dicksonia antarctica* and *Cyathea australis*, while tall tree ferns grow alongside one of the cast-iron spiral staircases which give access to the gallery. One of the delights there is to climb above the ferns, peer down into their gloriously spreading crowns and see the giant fronds unfurl like bishop's croziers, the size of a man's clenched fist.

Many other glasshouses stud the grounds of Kew: aroid; filmy fern; Australian; tropical water lily; tropical; succulent; fern and alpine.

The alpine house, opened in 1981, is a marvel of new technology, a pyramidal moated structure where water and ventilation are used to control the temperature. Around the central refrigerated bench is a miniature mountain landscape built from Sussex sandstone. On the bench Arctic alpines are grown experimentally, using banks of lights to fool plants into believing that the nights last only one hour.

The most recent development at Kew is the carefully landscaped Princess of Wales Conservatory sited near the rock garden. Although it is built very differently from the huge conservatory at Edinburgh, it follows the same internal philosophy of providing many different environments from mangrove swamp to desert terrain, and, like Edinburgh, it includes an underwater viewing aquarium.

Close by the new conservatory is the herbaceous ground, a fascinating area for the real plant enthusiast. This piece of land was used as a royal vegetable garden until it was incorporated into the main body of the garden in 1847. Now it has the feel of a traditional botanic garden, with plants arranged to demonstrate their relationships. Thousands of herbaceous perennials crowd the oblong beds, while overhead on the long pergola which covers the whole length of the central path, roses and clematis hang down in great swags.

1. Sir William Temple, *The Gardens of Epicurus, with other XVIIth Century Garden Essays*, Chatto and Windus, London, 1908.
2. William Chambers, *Gardens and Buildings at Kew*, 1763.
3. Ibid.
4. William Lawson, *A New Orchard and Garden*, London, 1676.

Sheffield Botanical Garden

Clarkehouse Road · Sheffield · South Yorkshire · S10 2LN
Tel. 0742-663115

*Open every day,
Mon–Sat 7.30 am–dusk,
Sun 10 am–dusk.*

Many cities acquired botanic gardens during the early part of the nineteenth century, the majority financed by private subscribers. Inevitably such subscribers were wealthy middle-class industrialists with enough leisure and money to indulge their keen interest in amateur horticulture. Designed by Robert Marnock, the Sheffield Botanic Garden opened to subscribers in 1836. Not only was it to be a beautiful open space at a time when the benefits of such spaces for city dwellers were being realized, but it was also to be a centre for scientific enquiry.

The science of gardening was developing apace. Many magazines devoted to the subject were being printed and gardening writers such as John Claudius Loudon were encouraging sound horticultural practice among their readers. One need only dip into Loudon's *Encyclopaedia of Gardening*, published in 1826, to savour the curious blend of science and philanthropy that was a characteristic of such writing. However, gardens such as Sheffield's stayed firmly shut to working men and women except on special and occasional open days.

In 1852 attempts were made at Sheffield to open the garden to the working classes at a reduced rate on one day each week. The move was defeated and greater accessibility to the garden for working people was not achieved until 1898, when severe economic problems forced the garden to be transferred from private subscription to the Town Trust. The Trust administered the garden until it was taken under the wing of the Sheffield Corporation in 1951.

Marnock was designing the garden at a time when Loudon's published ideas on gardening were gaining ground. The fashionable style of gardening was called 'Gardenesque', a curious mixture of formality and exuberant planting which included bedding out of tender exotics. The style owed much to the 'Picturesque', which embraced an irregular and often rugged natural beauty. This was a conscious reaction against the landscape gardens designed by Capability Brown during the previous century. Brown's landscapes floated the aristocrat's house in acres of grass, banishing all formality in the process. Brown's successor, Humphry Repton, reintroduced the formal terrace between house and landscape, thus creating an area of transition. No longer did the country house merge into the landscape which surrounded it. What the Gardenesque sought to do was to graft onto Repton's landscape style the very latest advances in horticulture.

The core of Marnock's design at Sheffield was to be the terrace, a subdued and elegant range of curvilinear conservatories occupying the higher ground to the north of the site. This range of conservatories formed the upper arm of a grand T-shape and the long walk which dropped away represented the formal transitional stage between order and contrived naturalness. Formal lawns displaying specimen trees abutted the long walk while beyond were the woodland areas, criss-crossed by a network of serpentine paths. These paths provided a variety of shifting perspectives for the visitor.

The overall design of the garden has changed little since Marnock laid down his original plan. The fountain at the end of the long walk has been replaced by a monument to those killed during the Crimean War, and the elaborate colonnades between the three pavilions have been replaced by less robust ones. However, Marnock would still recognize his garden today, even though the trees have grown to lush maturity. Indeed the garden is a mature, even bloated, Victorian original contained within its walled nineteen-acre (7.6-hectare), south-facing slope, a triangular site with its apex thrusting north, enjoying a mild micro-climate.

The main entrance to the garden, a honey-coloured neo-classical sandstone gateway, occupies this northern apex. Although somewhat pretentious, the gateway is second in grandeur only to Inigo Jones' entrance to the University of Oxford Botanic Garden. This apex is the highest point of the garden and from here the ground falls away to the south-west, gently at first but more dramatically where the main east and west lawns give way to the woodland garden. This peaty escarpment is held in check by terraces formed from logs and stones.

A short distance to the west of the entrance runs what Don Williams, the curator, calls the Paxton Terrace, three linked pavilions designed by Robert Marnock and Benjamin Broomhead Taylor, and influenced no doubt by contemporary conservatory design.

All three pavilions are built in the style of late eighteenth-century orangeries. Such orangeries would have had flat opaque roofs, but these pavilions have most beautiful curvilinear glass roofs, a distinct nineteenth-century characteristic that has given rise to speculation that Paxton was involved if not in their actual design then certainly as a consultant. The conservatories were described in the *Floriculture Magazine* of 1836: 'The entire line of frontage is 100 yards: but the extensive and beautiful structure itself is divided, it will be seen into five parts, the narrowest of which is twenty four feet in width. The terminating buildings as well as that in the Centre, forming the noble and commodious Greenhouses, the unglazed portions being built in front of rubbed stone, and having Corinthian pillars alternating with vertical sashes. Each of these houses is covered by a quadrangular glazed dome, constructed of metallic ribs connected and surmounted in the centre with an ornamental casting'.

*The Paxton Terrace, three linked pavilions which are now a conservatory, an
aquarium and an aviary.*

Only one of the three pavilions is still used as a conservatory. One is an aquarium and
the other an aviary from which macaws harangue the visitor with wolf-whistles. All three
were originally connected by lower covered structures. These have long since vanished
and only the aviary and the aquarium are now physically linked. A colonnade clad with
wisterias runs between them in front of which grow berberis, potentilla, rue, hebes,
yucca and abutilon.

On the lawns in front of the terraces are a number of island beds planted to
demonstrate a diversity of form, texture, foliage and flower. There can be found
ornamental kale, richly variegated purple and white, looking like a remnant from a
Victorian dress. Here too is *Euphorbia sikkimensis* and the yellow dogwood, *Cornus
stolonifera* 'Flaviramea'. Repeatedly cut back over the years, the latter now sends up a
thicket of vivid yellow stems which stand out in dramatic relief against a large clump of
pampas grass. Nearby grows that most wonderful of trees, *Prunus serrulata*. Its bark,
silken to the touch, is a rich, bright mahogany colour.

From the central aviary a path, 180 yards (164 metres) long, drops away dividing the
main east and west lawns. On one side is a long herbaceous border backed by a hedge of

clipped holly while on the other are trial beds. The lawns, mowed to a resilient springiness, have their edges trimmed to geometric precision and are studded with specimen trees, both broadleaf and conifer. As one approaches the Memorial which terminates the long walk one begins to sense the garden's almost claustrophobic profusion and its overblown maturity, infused with a Victorian concern for curved island beds laced together by serpentine paths.

Tucked almost out of sight down a slope to one side of the Memorial is a fractured and cracked fossil club moss. Three hundred million years or more old, it now sits upon the earth like a giant radiating claw.

To the south, beyond the Memorial, the ground drops suddenly away and nature, it seems, is given her head. This is the woodland garden, formerly an old quarry, where paths wind their way beneath mighty chestnut and beech trees underplanted with rhododendron, maple and azalea. In both spring and autumn it is a scene of unrivalled beauty. Lilies grow there early in the season while geraniums flower throughout the summer. In autumn the fallen leaves add a discreet and subtle luminosity to the narrow twisting paths. This is woodland proper, full of rich peaty smells and dense ground cover.

South of the woodland at the garden's lowest extremity is another area of fiery autumn colour, the sorbus lawn. To the west is the cedar lawn where conifers stand in precisely cut circles of earth and where broad turf paths weave between island beds. Rhododendron grows well there and so does that most English of trees, the yew. So generous has growth become that one's prospect of all but the most isolated of conifers is obscured.

The yews at Sheffield are a direct legacy of industrialization, because they were able to cope with atmospheric pollution during the previous century when other conifers fell by the wayside. And how versatile they are too, willing to shoot from old wood when cut hard back or making magnificent specimens when given their head. Evergreen yew, with its scaly bark flaking off to reveal red underwood and its intriguing light-absorbing canopy, has an alluring haunting presence.

Stories abound of the tree's poisonous properties. Gerard in his *Herball* wrote, 'The Yew tree . . . is of a venomous quality, and against mans nature. Dioscorides writeth, and generally all that heretofore have dealt in the facultie of Herbes, that the Yew tree is very venomous to be taken inwardly, and that if any doe sleepe under the shadow therefore it causeth sickness and oftentime death. Moreover, they say that the fruit thereof being eaten is not onely dangerous and deadly unto man, but if birds doe eat thereof it causeth them to cast their feathers, and many times to die. All which I dare boldly affirme is altogether untrue; for when I was young and went to schoole, divers of my schoole-fellows and likewise my self did eat our fils of the berries of this tree, and have not onely slept under the shadow thereof, but among the branches also, without and hurt at all'.

John Evelyn, the great seventeenth century arboriculturist who published his classic work on trees in 1664,[1] relates a story concerning the gardeners at the Medical Garden at

Pisa. When clipping the yew trees there, he says, they 'are not able to work above half an hour at a time, it makes their heads so ake'. And of the yew's almost legendary association with churchyards Evelyn writes: 'That we find it so universally planted in our churchyards, was, doubtless, from its being thought a symbol of immortality, the tree being so lasting, and always green'.

At Sheffield yews are everywhere; six clipped topiary specimens stand guard in front of the pavilions while lost in the woodland garden one magnificent specimen grows on a triangular island marking the meeting point of three paths.

The ascent through the shrubberies is pleasantly gentle, past beds of tree paeonies with hellebores running between their roots, and alongside island beds with a mixed planting of holly, rhododendron, beech and yew, shoulder to shoulder.

Above the shrubberies is one of the garden's most extraordinary features, the bear pit. This deep, perfectly round pit, open to the sky and set into the slope, offers evidence of the Victorian love for the bizarre and exotic. One observer recorded their childish fascination with the bear pit in 1856: 'There were two bears in the bear pit. In the centre

Only at the eastern extremity of the garden does the proximity of the city of Sheffield become apparent.

was an erection resembling a big tree with three platforms at intervals on branches. We children used to aim chunks of bread or buns onto the platforms to tempt the bears. When the bears climbed to the higher platforms the trees swayed and the children's excitement was great when it came near the railings.'[2] These railings are still there today, though the bears have long since gone.

The rose garden which the Ordnance Survey map for 1856 delineates as an ornate romantic knot is now a parterre of rigid formality contained within a tall hedge of clipped yew.

At each end of the rose garden is a concave space reminiscent of William Kent's exedra at Chiswick House in London. But instead of Chiswick's gathering of sculpted states-men, Sheffield's exedra boasts an array of blooms. Compared to the feeling engendered by the profuse planting in the rest of the garden, the feeling here is one of an enclave of sobriety. Beyond the yew hedge, beech, oak, conifers and ornamentals press in or raise their crowns as though to peer in at the delicate confined blooms.

Above the bear pit, is that most special of exotic trees, the liquidambar, whose truncated leaves cover the branches with vivid autumn colour until well into the winter, when all of a sudden they seem to fall as one. Next comes the heath garden, wild with colour in autumn and in spring a multitude of crocus, tulip, magnolia and lily.

The aptly named smoke tree, *Cotinus coggygria*, seems adrift in a shifting cloud of bluey mauve in early winter. *Pinus coulteri* is there too, a handsome tree from California with delicate needles growing in long clusters of three. The spiky cones, which grow to twelve inches (thirty centimetres) or more in length and weigh up to five pounds (2.2 kilos), have given the tree its common name, the big cone pine.

Only at this eastern extremity of the garden does one become aware of the proximity of the city of Sheffield. Having clambered almost unaware back up the slope one stands on a level with the Paxton Terrace. The glass roofs break the skyline, lifting themselves above the low canopy of ornamental trees, while beyond them, just a mile and a quarter (two kilometres) away, office blocks appear against the sky.

A semi-woodland walk, the ground covered in thick mats of ivy, runs behind the terrace. The view through the colonnade is reminiscent of the old painterly trick of alternating areas of light and shade to create depth. Glimpses of the light terrace and lawns beyond break through the dark confines of the colonnade. On the other side of the Clarkehouse Road entrance is the Victorian garden, an area of bedding out in summer, a patch of bare brown earth in winter. Beyond this is a rather mean rock garden, then the delightful nature garden, a pocket-handkerchief size space of unashamedly didactic intent.

The tiny nature garden, run by Sheffield Friends of the Earth, is designed and planted specifically with plants that attract wildlife. Many old country favourites grow here: corn cockle, corn marigold and cornflower, all of which are becoming rare in the wild. Teasels, chicory and cow parsley flourish; the latter dries to a seed-hung skeleton of majestic elegance. The biennial Aaron's rod, *Verbascum thapsus* is also there, with buttercup-yellow flowers encrusting a tall stalk that flares from a rosette of soft grey leaves.

In this tiny patch are many flowers which are readily available from commercial growers even though rare in the wild. The point is made on an adjacent display board that in England there are 9,884,000 acres (4,000,000 hectares) of private garden and if everyone planted for conservation, wildlife would have a better chance of survival.

The nature garden highlights the didactic value of botanical gardens such as Sheffield where teaching is achieved by example. There may be no herbarium or research facilities at Sheffield and it may seem closer to an ornate public park than the generally accepted notion of a botanic garden, but a sound horticultural function is still maintained. The garden is a venue for a number of local societies and in 1985 alone 16,000 specimens were supplied to local schools for horticultural research.

1. John Evelyn, *Sylva*, 1664.
2. Jan Carder, *The Sheffield Botanical and Horticultural Society Gardens*, Dec. 1982.

Southampton University
Botanic Garden

Biology Department, · Building 44 · University of Southampton · Southampton · SO9 5NH
Tel. 0703 559122 extn. 2038

Open Mon–Fri,
9 am–5 pm;
closed Christmas week, Easter
week and Bank Holidays.
No entrance fee.

Southampton University has a modern campus built on a gentle slope at the edge of the city. A natural stream cuts the site, and on either side the lawns are neatly tonsured. A beautiful selection of plants, including azaleas, hostas and ornamental rhubarb, flank the stream and a perfectly kept rock garden fills a steep slope. By the side of the biology building a great thicket of bamboo hides the stream from view before it drops out of sight down two vertical culverts. From here the botanic garden itself is still hidden from view, behind the biology and the students' union buildings. The visitor who cautiously follows the path around those buildings is confronted by an inconspicuous gate flanked by two pencil thin incense cedars, *Calocedrus decurrens*. This is the entrance to the garden. It has about it the air of a municipal allotment, efficiently business-like and with little regard given to aesthetics.

Indeed, this sums up the atmosphere that prevails within this garden, where few concessions have been made to landscaping the grounds. However the two-acre (.8-hectare) site, squeezed in between the university buildings and the adjacent residential houses – both of which are obscured by mature trees – enjoys a natural beauty.

The garden occupies a valley aligned on a north-south axis. The stream which has re-emerged meanders along the valley bottom providing a home for a lush planting of water-loving specimens. Primulas, hostas, rodgersias, *Rheum palmatum*, masses of sweet cicely, and the elegant *Miscanthus sinensis* 'Variegatus', its strap-shaped foliage slashed with yellow, grow along the banks. At the far end of the garden where the stream runs away from the site are ancient gunneras, their huge extravagant leaves spreading beneath a towering oak and a young *Metasequoia glyptostroboides*.

On the east side of the valley are grass terraces, each filled by a long flower bed in which plants required for study by the students are grown. Bloody cranesbill, *Geranium sanguineum*, and *Asarum europaeum* are packed hard up against a large clump of false Solomon's seal, *Smilacina racemosa*. There is a bed devoted to the *Liliaceae* family, including the beautiful *Asphodeline lutea*, its vivid yellow flowers carried on a tall shaft

The site of Southampton University Botanic Garden enjoys a natural beauty.

above glaucous foliage, and *Hosta sieboldiana*. *Papaver nudicaule*, the Iceland poppy, seeds itself freely, its sun-orange and egg-yolk yellow flowers providing vivid splashes of colour on the slopes.

Above the terraces the slope rises sharply to a woodland border where shrubs and trees mix; *Sorbus prattii*, *Osmanthus delavayi*, masses of foxglove, *Acer griseum* and *Carpinus betulus* grow from mats of springy moss.

Further along the slope the grass area opens out on to a bank of shrubs. Prominent among these is *Pieris forrestii*, its new foliage in spring brilliantly red and pink and its flowers carried like great swags of lily-of-the-valley. Three other plants there catch one's eye: a mature fig; a rampant *Actinidia chinensis*, with lush young foliage; and the weed-like perennial *Amsonia tabernaemontana*, the prolific qualities of which seem to be threatening plants on all sides. Only a mature and dense clump of epimedium appears to be holding its own against this plant's onslaught.

Slightly below the epimedium is the artificial pond with royal ferns, rushes and an old *Acer palmatum* 'Atropurpureum' reflecting its delicate foliage in the shallow water. At one end water dribbles from the pond's fractured edge and trickles down past a vigorous clump of *Trillium grandiflorum*.

Most of the western slope of the valley is occupied by two glasshouses, only one of which is open to the public. In this relatively small glasshouse exotics are crowded together. The Canary palm rubs shoulders with abutilon; a dragon tree, *Dracaena draco*, tangles with the feathery mimosa, *Acacia dealbata*, and clerodendrons pierce the gloom with their brilliant flowers. For sheer audacity of bloom, however, there can be few plants to beat *Aristolochia elegans*. This climber scrambles high along the side of the glasshouse and hangs down its bizarre flowers, short, cream-coloured bulbs attached to wide, fan-shaped hoods veined with purple and mauve and somehow reminiscent of a medieval damask tapestry.

One end of this glasshouse is sectioned off and the temperature and humidity kept high. Economic plants grow there, the most impressive of which is a banana hung with ripening fruits. The lee of the glasshouse is home for a large *Crinodendron hookerianum*, its bright-red flowers hanging from the branches like miniature lanterns.

Between the two glasshouses is a fascinating herb garden where in tiny square beds a range of herbs grow in a similar way to those grown by monks centuries ago.

Above the herb beds the ground rises to a small pinetum, where the stone pine, *Pinus pinea*, can be found, the rarest plant in the garden and possibly the largest specimen in the British Isles. This spreading pine has an attractive feathery appearance.

The other trees of note in the garden are a beautiful *Eucalyptus perriniana* with tawny bronze bark and an aristocratic *E. niphophila*. The bark of the latter is a silver-grey to blue-white colour as though painted by a water-colourist.

St Andrews University Botanic Garden

The Canongate · St Andrews · Scotland
Tel. 0334-76161 extn. 8448

Open daily
April and Oct, 10 am–4 pm,
May–Sept 10 am–7 pm,
Nov–March Mon–Fri,
10 am–4 pm;
glasshouses 2–4 pm
Mon–Thurs,
10 am–3.15 pm Fri.
Entrance fee.

The harsh reality of economics is a continual threat to all botanic gardens. Their scientific benefits are all too easily forgotten and their amenity value denigrated. St Andrews is not the first to be faced with the problem of having to raise funds privately if it is to survive in full, and not be sold off for residential development. The garden's curator, Bob Mitchell, a man of infectious enthusiasm and boundless energy, seems equal to the task. He and a handful of staff have been responsible for the creation of the garden since its move to the present site just over twenty years ago.

Although the University of St Andrews, dating from 1411, is the oldest in Scotland, the botanic garden can only trace its roots back to 1889 when a Dr John Wilson laid out a quarter-acre (.1-hectare) plot in a walled garden at St Mary's College. From then, expansion of the garden went hand in hand with expansion of the University until in 1960 the new eighteen-and-a-half (7-hectare) site at The Canongate was acquired.

The split-level site with a steep but short north-facing slope was fields, pasture and a small woodland planted on the lower level of what was once a quarry. The transformation has been quite remarkable, with major trees now affording protection to the north and a belt of conifers subduing the winds that blow across from Norway.

Throughout the garden, conifers, including Scots pine and Austrian pine, play a crucial role, protecting the numerous borders. They also create different levels of interest; tall trees protect tender shrubs which, in turn, nurture ground cover.

The garden possesses a seductive beauty and it is easy to see its main role as being that of amenity rather than scientific. The curvilinear beds of trees and shrubs interspersed with luxurious herbaceous perennial planting, the gravel and grass paths snaking around island beds are as fine an example of good gardening as can be found anywhere.

There is a long, deep herbaceous border, in front of which are the semi-circular order beds. This part of the garden has a dignified, quiet charm, but the romance of the herbaceous border backed by its neat clipped beech hedge and the precision of these scientifically arranged order beds do nothing to prepare the visitor for the drama of the

peat slopes which drop precipitously to the north. The two adjacent areas, neither of which encroaches upon the other's authority, are a perfect example of the successful juxtaposition of order and controlled abandon.

The central features of St Andrews are the peat terraces and the adjacent rock garden crafted from the cool north-facing slope. There are plants in ecological groupings, in a terrain that suggests everything from mountain to wet fen, limestone pavement to coarse scree. On these slopes everything is turbulent and natural as plants grow into one another with happy familiarity, nestling under young oak and Scots pine, the latter grown from seeds collected from remnants of the Caledonian Forest.

There are dozens of rhododendrons there, mainly dwarf varieties. In the spring their flowers sparkle among the dense foliage, and in the autumn their leathery leaves flash grey and silver and green in the sun while ornamental knotweed, *Polygonum affine*, binds the lower terraces with a glorious pink slash. Primula, meconopsis, *Helleborus niger*, saxifrage and gentian thrust their roots down into the rich peaty soil.

The sensitive fern, *Onoclea sensibilis* grows there, its fertile spores clustered like tiny jade spheres around the upright stems, green at first and then brown during the winter, while the sterile fronds are the first things in the garden to be nipped by the autumn frosts. The giant lily, *Cardiocrinum giganteum* can be found here too, bold and rude, its magnificent flowers eventually giving way to huge bloated seed pods.

At the centre of the garden is the rock garden. Compared to that of the Royal Botanic Gardens in Edinburgh, it is profusely wild and small, yet it has a more satisfying look, with its great slabs of whinstone and basalt emerging naturally from the slope. There are masses of saxifrage and penstemon, laced by the rampant trailing stems of perennial nasturtium, *Tropaeolum polyphyllum*, and mats of purple edraianthus and *Euphorbia capitulata*. There is *Mentha requienii* from Corsica which releases an intoxicating fragrance when crushed underfoot, and *Juniperus procumbens*, which observes with gentle fidelity the contours of the rocks.

Some native Scottish gems can be found too: *Dryas octopetala*, for example, named after the wood nymph and now extremely rare in its native habitat in the west of Scotland. *Silene acaulis*, the moss campion, fares better, growing into dense tough tufts between cracks in the rock. Of this plant, John Parkinson wrote in his *Theatrum Botanicum* of 1640, '. . . it hath no great sent to commend it, but onely the beauty of the verdure, and blush, so thicke intermixt like a wrought carpet to please the sight'.

It is a marvellous experience to sit above the rock garden on a bright sunny, slightly windy day and watch the light travel across the boulders and plants. The scene changes each second, while the trees at one's back hiss and bend in the wind.

The rock garden, one of the central features of St Andrews.

The glasshouses are interesting, but more interesting are the plants such as *Echium wildfretii* that grow huge and vulgar in the mild micro-climate created by the glasshouses.

Most intriguing, and sadly tucked out of sight of the casual visitor, is the huge deep-freeze cabinet deep within the glasshouse complex. There myriad blooms, picked from the garden when at their best, have their perfection locked in a timeless chill. When term begins they will be rescued from their icy chamber by students who will struggle with their own taxonomy. Until then, these dazzling frozen colours, fresh and frosted, wait to be exhumed from the dark.

University College of Swansea Botanic Garden

Singleton Park · Swansea · Wales · SA2 8PP
Tel. 0792-295386

Open Mon–Thurs
9 am– 4.30 pm,
Fri 9 am–4 pm
except from Christmas Eve
until 2 Jan (or next
opening day).

The most novel experiment I have encountered in years of travelling around gardens is to be found tucked away in one corner of the University College of Swansea Botanic Garden. A low steel barrier surrounds a circular area approximately twenty feet (six metres) across. Slung over it is a fine mesh net, the shape of which resembles a sagging bedouin tent. A prominent red warning sign guards the entrance: 'No Admittance, Danger,' it reads.

Within the tent an experiment on adders' habitats is taking place. So successful has it been that the snakes, collected from the wild, have bred and seem perfectly at home. It is one of those slightly eccentric flights of fancy that give visitors to such gardens an extra frisson of surprise and pleasure.

Swansea has other attractions, too. The twenty-seven-year-old garden is small, two and a half acres (1-hectare) at the most, forming an integral part of the university campus. It is in effect an island of green with buildings to one side and the edge of the town to the other. The lush planting verges on the unkempt and succeeds in acting as a foil to the modern university buildings against which it is pressed.

The main body of the garden, which slopes gently to the south (the slope accommodated by low terraces) is rectilinear and rigidly geometric. However, the formality is softened and even obscured by the planting and by the many mature trees on site.

Dividing the garden is a long pergola which skirts a central courtyard. Where the pergola touches the edge of a terrace, and steps descend, a Lutyens feel is evoked. The brick piers of the pergola support a host of roses, clematis and vines, and the pink limbs of an old *Vitis coignetiae* are contorted into a variety of fantastic shapes.

Near the courtyard a white wisteria displays dazzling blooms across the spans of the pergola and beyond, high into an adjacent tree. Fragrant tobacco plants, *Nicotiana affinia*, fill the beds and to the north, on the upper terrace, the scent from a rose garden swamps the air. Also growing here is a huge tulip tree, *Liriodendron tulipifera*, and a clump of oriental poppies adds a vivid, blood-red splash of colour.

Wild flowers and weeds have seeded themselves in most of the experimental beds,

Part of the long pergola at Swansea Botanic Garden.

though whether intentionally or through neglect is unclear. However, one or two of the beds deserve special attention. Worth noting on the northern terrace is a vast *Griselinia littoralis* and a delightful bed of geraniums pierced by tall lupins and lilies.

On the lower south terrace the emphasis is more on foliage plants, with trees and shrubs happily intermingled. *Populus lasiocarpa* with its long heart-shaped leaves grows near a *Paulownia lilacina*; *Cotinus coggygria* grows near a sprawling *Cotoneaster lactea*, and yuccas and phormiums flower side by side.

To the east of the pergola is a tiny moated island, the moat being little more than a shallow serpentine rill. To one side is a rock garden with great slabs of stone and an area of shallow scree. Maples, primulas, dwarf conifers, Welsh poppies and geraniums all grow on the island. Above the water damsel flies dart.

A narrow road separates the island from a larger pool and terrace. Large phormiums and a Chusan palm, *Trachycarpus fortunei*, give the terrace a Mediterranean feel. To one side is a tiny pinetum, to the other a marshy extension to the woodland garden, which occupies a long strip of undulating land along the north-west edge of the garden and thus forms a solid bank of green between town and gown.

Wakehurst Place

Ardingly · Haywards Heath · West Sussex · RH17 6TN
Tel. 0444-892701

Open every day except
Christmas Day and New Year's
Day.
10 am–4 pm (winter),
10 am–7 pm (summer).
Entrance fee.

Wakehurst Place, a splendid Elizabethan mansion once owned by Sir Edward Culpeper, is Kew in the country: an annexe of the famous Royal Botanic Gardens. There, in the fresh, unpolluted atmosphere of Sussex, rhododendrons and conifers grow to luxuriant maturity.

The major botanical development at Wakehurst took place after the estate was acquired by Gerald Loder in 1903. For the next thirty-three years Loder, excited and stimulated by the discoveries pouring in from the Himalayas, packed masses of rhododendron, magnolias and conifers on the slopes and in the valleys of the estate. When he died in 1936 the estate was bought by Sir Henry Price who maintained and extended Loder's collection. On Price's death in 1963, the estate was bequeathed to the National Trust which, in turn, made it available to the authorities at Kew.

The grounds of Wakehurst are typical of many turn-of-the-century collectors' gardens, where overall design took second place to a setting in which to display prized specimens. Wakehurst triumphs, however, because the exotics have merged naturally into the landscape which is both disparate and rugged.

The estate is made up of two valleys which meet at the man-made Westwood Lake and lie upon the land like the open jawbone of a pre-historic monster. It is set down on an approximate east-west axis. The valley sides are by turns steep and gentle; a precipitous ravine at one part drops to a tumbling stream, and elsewhere a grassy shelf describes a gentle curve.

The mansion occupies a plateau at the extremity of the lower jaw. The grounds close to the house are formal but lawns give way to irregular beds of heather and shrubs and then progress naturally to pinetum and woodland where the planting becomes vigorous and abandoned. South-east of the house is a small ornamental pond, the chief purpose of which is to reflect the lightness of the design and the warm colours of the sandstone.

The pond also provides a setting for a rock garden and an extensive planting of conifer and rhododendron, including the scarlet-flowered *R. barbatum* and *R. x shilsonii*. Close by is the beautiful, white-flowered *Magnolia wilsonii*, which was introduced by the

plant-hunter, E. H. Wilson in 1908. Ground cover plants are also skilfully used here. *Geranium procurrens* and *Polygonum vacciniifolium* scramble among rocky terraces and *Pachysandra terminalis* makes steady progress around such exotics as the Chusan palm, the only palm which makes any pretence of being at all hardy in Britain, and the magnificent bamboo, *Arundinaria murieliae*. The small pond is also home for some fine clumps of royal fern, *Osmunda regalis*, whose fertile fronds rise like tight curled cinnamon sticks in mid-summer.

From this small and rather refined pond a water course tumbles over rocky outcrops to form further pools among sylvan glades before meandering off along the valley. This area, known as The Slips, is where the drama of the garden begins. Among these ponds grow many British natives, including the narrow-leaved lesser reedmace, *Typha angusti-folia*; sweet flag, *Acorus calamus*; and the flowering rush, *Butomus umbellatus*.

Beyond the pools the water empties into a ravine. The footpath divides and shins off along the upper edge of Westwood Valley where it teases the visitor with the most dramatic and theatrical views. On the slopes of this deep gash hundreds of species of rhododendron and azalea grow beneath the canopies of ancient beech trees through which the light pours in long broad shafts. There *Rhododendron succothii, R. delavayi, R. falconeri, R. sinogrande* and *R. barbatum* flower with dazzling blooms in early spring. Thick bracken at times seems set to overwhelm the slopes. Trees include the Chusan palm; the tall, elegant western hemlock, *Tsuga heterophylla*; *Magnolia campbellii*, which is claimed to be the tallest specimen in the country; and that most beautiful of trees, *Davidia involucrata*, hung with translucent tissue-paper bracts early in the year.

The rediscovery of this tree by E. H. Wilson in China in 1899 is a story in itself. John Veitch, the successful nurseryman, financed the two-year expedition after having seen dried specimens of the tree sent back to Kew by Augustine Henry, the customs and medical officer then based at Ichang on the Yangtze River. Wilson's journey was fraught with danger. He had to spend many weeks in China waiting for a volunteer to take him up the Red River where murders of Europeans were commonplace. Then, when the volunteer did arrive, he proved to be an opium addict who on more than one occasion almost wrecked their boat on rocks.

Eventually Wilson reached Henry who told him where, twelve years earlier, he had found *Davidia involucrata*. The tireless Wilson set out only to find that the tree had been cut down and the timber used in the construction of a native's hut. Although thwarted on this occasion, Wilson went on to find other specimens and, after a tense and expectant wait for the seeds to ripen, was able to harvest them and send them back to England. It is

The pleasaunce, a square of clipped yew hedges within the formal walled garden at Wakehurst Place.

a nice thought that perhaps the plant at Wakehurst was one of the original seedlings brought on at Veitch's nursery sometime in 1903, the year Loder bought the estate.

In spring Westwood Valley is alive with the sugary fresh blooms of rhododendrons and azaleas, and lady's smock, *Cardamine pratensis*, and masses of common spotted orchid thrive at ground level. In autumn the beech leaves fall in luminous golden clouds and the bracken adopts a russet colour.

Further along the valley, the ravine widens into the Himalayan glade and Rock View, where a dramatic feature has been created from the natural outcrops of sandstone. The precipitous crags are carpeted in plants that would grow at an altitude of 10,000 feet (3,048 metres). There are dense thickets of low-growing *Berberis wilsoniae*, and *Polygonum affine* adds colour in summer and autumn. The rocks are stained with lichens, and ferns seek out what often seem to be inaccessible damp cracks.

Rock View allows fine views down the widening valley to Westwood Lake, which glitters through the trees. At one's back the twenty-acre (eight-hectare) pinetum laid out in 1914 rises up a gentle slope towards the house.

It is a mistake to head back through the pinetum; instead, go beyond the prettiness surrounding the house and the obvious spectacle of the Himalayan glade to the more subdued charms of Horsebridge Woods. There are masses of bamboo, some low-growing and advancing in carpets through the woodland, others tall and growing into elegant feathery clumps.

The actual planting in Horsebridge Wood is quite thin, and allows thousands of bluebells to grow close to the conifers. *Sequoia sempervirens* reaches for the sky while the tallest tree on the estate, *Abies grandis*, is allowed plenty of space to breathe. A very tall monkey puzzle, the most bizarre of trees, is attractively juxtaposed next to what has become known as our Christmas tree, *Picea abies*.

The path eventually leads to Rock Walk, a track which runs beneath outcrops of High Weald sandstone for three quarters of a mile (1.2 kilometres). This is the home of countless yew trees, which find the sharp drainage and dank humidity to their liking. Many of the yews seem to spring from earthless fissures and send out twisted and contorted roots across the rocky surfaces. The dense canopies create gloomy deep tunnels in which great swags of ivy coat the rocks, some of which are stained a rich emerald green.

Beyond Rock Walk is Bloomer's valley, a long curving sweep of grass set like a shelf into the valley slope. At the far end a log cabin dominates the prospect.

The track through the valley passes conifer and deciduous trees until, in Horsebridge

Wakehurst Place, where fine specimens have merged naturally into the rugged landscape.

Wood, it splits. One path drops down to Westwood Lake, which has a constant greenness as it reflects the tangled vegetation of the surrounding slopes. On the far side of the lake the land climbs sharply up to the southern edge of Westwood Valley and commands a beautiful view of the Loder Valley Reserve and the Ardingly Reservoir.

On these slopes as elsewhere the planting is mixed, with beech, oak and conifer around thickets of rhododendrons. The undergrowth can be quite dense and it is not unusual to surprise pheasants which crash away in alarm. This point high above the Himalayan glade marks the end of the wild and exuberant woodland and a return to the formality and order of the shrub borders and heath garden.

The heath garden leads by beds of crinums and agapanthus to the walled garden, which asserts a further degree of formality. Within the walls are two enclosures. One, the pleasaunce, is a square of precisely clipped yew hedges conceived with an architectural grandeur. Safe within the hedges are cruciform flower beds set in grass. At the central axis a fountain plays. It is an eloquent renaissance scene, both calming and contemplative, which contains the seeds of all the splendour gone before.

Hard by the pleasaunce is the Henry Price Memorial Garden, constructed in 1975. Here is an assembly of grey-leaved and silver herbaceous plants from which all brittle and harsh colour is banished. Everything within this secluded garden, from herbs to flowering annuals, has been chosen to achieve a shimmering confection of gentle colour.

Westonbirt Arboretum

Tetbury · Gloucestershire · GL8 8QS
Tel. 0666-88220

Open every day,
10 am–8 pm
(or sunset if earlier).
No entrance fee.

Westonbirt Arboretum in Gloucestershire is a tribute to Victorian acquisitiveness and to the competition between wealthy industrialists, who vied with each other to get hold of the latest plants sent back to Great Britain by plant-hunters such as David Douglas. The period was also characterized by a great thirst for sound gardening knowledge. This was slaked by writers such as John Claudius Loudon, whose book, *Encyclopaedia of Gardening*, published in 1822, was packed with useful information.

A Loudon-inspired garden included a bit of everything, occasionally reduced to ostentatious – some would say garish – display. But if Loudon sowed the early seeds of such excess with a style which came to be known as 'Gardenesque', he also had a fervent attachment to arboriculture. His book, *Arboretum et Fruticetum Britannicum*, published in 1838, was a precise, eight-volume exposition on trees and shrubs of the British Isles. In many ways the book echoes the earlier classic on arboriculture, John Evelyn's *Sylva*, published in 1664. Loudon, however, also embraced the new species that had been introduced and his work reflected the love of trees that persisted through the Victorian age. He exhorted his readers to go out and plant trees and expressed the hope that the book would introduce 'a greater variety of trees and shrubs (into the) plantations and pleasure grounds . . . among gentlemen of landed property.'

One gentleman, who by 1838 was already passionately involved with trees, was Robert Stayner Holford. The soil close to his house at Westonbirt near Tetbury was a rather poor limestone overlaying rock, but just half a mile (.8 kilometre) away was sandy loam covering an area of some 114 acres (46 hectares). It was there, on the open arable pasture to the west of his estate, that in 1826 he set to work planting his acquisitions from around the globe. At the same time he began to establish long elegant rides in an adjacent woodland, founding what is today the finest and most extensive arboretum in Britain.

Robert Holford's passion for new and unusual species was shared by his son, Sir George Holford, and together they supported many plant-hunting expeditions. It was Sir George, in fact, who really developed the arboretum's range of spring and autumn

colour, with lavish plantings of rhododendron and maple, while helping to extend the original arboretum across a shallow insignificant valley into Silk Wood. Eventually, in 1926 the estate passed into the hands of Sir George's nephew, the Fourth Earl of Morley; thirty years later it was handed on to the Forestry Commission.

The 500-acre (202-hectare) site is a mass of oak, chestnut, beech and larch, planted in blocks which in turn shelter more tender exotics. Broad rides separate these blocks and sinuous paths help to link the many glades. The emphasis is very much on the juxtaposition of specimen trees. There is no water or landscaped feature to add variety.

To Robert Holford we owe the large number of skilfully sited conifers such as the Douglas fir, *Pseudotsuga menziesii*, which stand near the oldest part of the arboretum, and the tall Wellingtonias, *Sequoiadendrum giganteum*, which stand on the Mitchell drive. But Sir George Holford is responsible for what has become Westonbirt's finest feature – the autumn colouring trees, of which the most eye-catching are the maples

Above *The impressive main gates to Westonbirt.*

Left *Autumn colour in Acer glade at Westonbirt Arboretum.*

which flank the path stretching almost 200 yards (182 metres) in Acer glade. This mixture of planting, coupled with great blocks of trees cut by long straight rides or stitched together by serpentine paths, is a characteristic of Westonbirt.

The lack of overall structure or formality in the planting encourages the visitor to wander at will and to drift from oak to beech. There is something to see at almost any season of the year. In January, the woods are spangled with witch hazel and *Parrotia persica*; in spring Silk Wood is carpeted with bluebells, and rhododendrons add their own distinctive bright colours; in late autumn, after the magic of the Japanese maples is past, fallen beech leaves add a golden luminosity to the paths.

Winkworth Arboretum

*on the B2130 between Godalming
and Hascombe · Surrey
Tel. 0483-67430*

*Open every day,
dawn to dusk.
No entrance fee.*

There can be few landscapes in the south-east of England that match the grandeur of Winkworth Arboretum – ninety-six acres (thirty-nine hectares) of woodland and lakes, given to the National Trust in 1952 by Dr Wilfred Fox. Fox was an amateur tree enthusiast and founder of the Roads Beautifying Association, a little known organization pledged to planting trees along the verges of roads that were encroaching into Surrey in the 1930s.

When he bought the land from the actress Beatrice Lillie in 1935 it was suffering from years of neglect. Bramble, hazel and bracken had been allowed unrestricted access and had overrun the steep valley slopes. The redeeming feature of the site was this valley, with its two long artificial lakes. Rowe's Flashe had been flooded in 1896, Phillimore Lake much earlier. These lakes create a scene of outstanding natural beauty and give the valley an aura of timeless enchantment.

In 1935, with very little help, Fox set about clearing the dense undergrowth, wisely leaving untouched the remnants of an old oak woodland which abuts the east shores of Phillimore Lake. This area, full of wood anemone and bluebells in the spring, has all the quiet solemnity of an old English woodland. Fox planted many fine deciduous trees, including sorbus, maple and cotinus. They provide a magnificent display of autumn colour and it is for this that the arboretum is mainly known. In his old age Fox created what he called the Carlotta Glade, since renamed the Azalea Steps, a steep grassy slope flanked with dwarf evergreen azaleas against a backdrop of maples. In spring this zig-zagging climb is a bright slash of colour against the fresh green slopes.

In 1952, at the age of seventy-seven Fox gave most of his estate to the National Trust, followed by an additional thirty-five acres (fourteen hectares) five years later. He died in 1962 and today this fine collection of trees has grown to gentle maturity, even receiving from time to time rare seedlings from the Royal Botanic Gardens at Kew.

The charm of Winkworth is its air of controlled abandon, a far cry from the neat clipped hedges and tonsured trees that characterize so many botanic gardens. The site is essentially a steep hillside dropping down to a valley aligned on a north-south axis. To the west the valley slopes, thickly planted with trees, climb to a plateau which leads to the

main car park. To the east the more gentle slopes are given over to farmland. The arboretum is beautiful in both spring and autumn.

'Of all the Cone trees onely the Larch is found to be without leaves in the Winter,' wrote John Gerard in his *Herball* of 1636. True enough, but what he failed to say was that a larch plantation in autumn, when the falling needles drift in great yellow clouds, is the most exhilarating sight, albeit touched with melancholy.

At Winkworth one can do precisely that, for just inside the main entrance at the western extremity of the arboretum stands a larch plantation, an acre (.4 hectare) of straight backed trees, their roots lost among bracken, their lofty crowns stretching up to the sky. These trees also serve as a massive wind-break, and the dense three feet (.9 metre) high bracken turns from fresh green in spring to tones of amber and rust, crinkly and brittle, in autumn.

On one side of the larch plantation stands a string of tall old Scots pine, their higher trunks splashed orange and blurred among the upper foliage. To the south of the plantation runs Perrydean Walk, flanked by small mahonias, many of which are dressed in yellow flowers, and a great bank of holly sprayed and stippled with scarlet berries. Beyond the holly is a deciduous woodland and the winter garden, the latter a collection of witch hazel, viburnum and ornamental cherry. Nearby stands a plantation of coppiced sweet chestnut. Every few years, when six feet (1.8 metres) in height, they are cut and used for the local manufacture of walking sticks. This almost impenetrable thicket is towered over by mature chestnuts and tall, whip-like silver birch.

Perrydean Walk strikes off into the heart of Winkworth, past the larch plantation and the summer garden, a collection of summer-flowering trees and shrubs, such as eucryphia, clethra, hoheria and *Hydrangea paniculata*. The most beautiful, however, is a group of *Stewartia pseudocamellia* from Japan; its seductive bark is flaky and pinkish orange, touched with patches of weathered green.

As the land begins to drop toward the precipitous valley-side there is a memorial to Dr Fox. A circle of benches appears suddenly among the trees like a Druid's temple. Two tall *Eucryphia* x *nymansensis* stand guard, one on either side, like soldiers silently watching the circle, while a young sweetly flowered *Viburnum tinus* sends roots into the sandy soil.

A tall purple beech, *Fagus sylvatica* 'Riversii', with leaves like thin beaten bronze, grows nearby but the best memorial to Fox is the tall whitebeam, *Sorbus* 'Wilfred Fox' which bears his name. Its autumn leaves are deep mahogany brown above, fawn underneath, rimmed with vivid yellow.

Further down the wooded slopes are masses of holly and magnolia, the latter a blaze of colour in spring, yet drab and somehow structureless for the rest of the year. Only

The Azalea Steps at Winkworth.

gradually from here does the view across the valley begin to open up until quite suddenly the slope drops away and the sky appears above one's head.

The drop down the slopes is dramatic. Spread out below are the canopies of many tall forest trees, including mature oaks, and wispy birches, which allow a fleeting glimpse of Rowe's Flashe, a great sheet of battleship-grey water. The other side of the valley gradually ascends, its agricultural parameters defined by ancient lush hedgerows.

The nearer, east-facing slope is densely clad with trees chosen by Fox for their autumn colour. Many maples and stewartias grow along the top edge of this valley, the latter with next year's plump buds, fastigiate and upward pointed, ready to displace the fading autumn foliage.

Views from the south escarpment are breath-taking, stretching along the whole length of the valley and on to the horizon far beyond. In autumn, the valley slopes can be seen in all their glory, vivid reds, nut browns and fading green flushed with pink, mauve and grey as the light shifts among the trees. Conifers such as *Cedrus atlantica* 'Glauca' contrast with the deciduous shamble, and the filigree branches of some bare trees throw spindly shadows when caught in shafts of sunlight; it is haunting and incredibly beautiful.

From the south escarpment, a path zig-zags down through the flattened bracken of Sorbus hill to a level area called The Bowl where tall oaks stand beside dark evergreen holly trees. There is a grove of *Acer capillipes*, with stunning orange-red leaves, and *Acer palmatum*, its few remaining leaves coloured like French chocolate with a slight white bloom. Nearby is a giant beech, its long branches drooping across the path. Alongside Rowe's Flashe stand a row of *Cedrus atlantica* 'Glauca' and another group of larch.

At the head of Rowe's Flashe, tucked behind a great swag of bamboo, sits a boathouse fashioned like a Canadian log cabin. Just a short distance away at the foot of the Azalea Steps is a tiny chalet almost overwhelmed by the weight of an enormous wisteria. A *Davidia involucrata* and a dwarf Rhododendron grow nearby, the latter's bonsai-like branches bearded with lichen.

Rowe's Flashe curves into a quiet lagoon where trees dip their branches to the water's surface. Offshore is a tiny island completely covered by bamboo. Only at this point does one become aware of the second lake, stretching away to the north beyond a boggy area, which in spring is a mass of marsh marigold and lysichiton.

The walk along the west bank of Phillimore Lake passes through the old oak woodland. This is a still and quiet place where few people penetrate and where the silence is punctuated by the echoing calls of the many birds that live here. The light beneath the spreading oaks in this wild place can be gloomy. The lake is still, its edges swampy. The northern extremity of the arboretum is marked by another boathouse, its watery entrance open like a dark gaping mouth, its reflection floating on the smooth glassy surface of the water.

Younger Botanic Garden

By Dunoon · Benmore · Argyll · Scotland
Tel. 0369-6261

In 1928, Harry George Younger gave to the nation his 10,000-acre (4,047-hectare) estate at Benmore in Argyllshire. Much of the estate had been heavily planted by James Duncan, a previous owner; between 1870–83 he planted a staggering 6,480,000 trees over 1,622 acres (656 hectares), creating a forest where once was a bleak valley. Many of these conifers have now reached magnificent proportions. The most striking are the Californian big trees, *Sequoiadendron giganteum*, which form a 300-yard (274-metre) long avenue near the main gate. But these, now over one hundred years old and in excess of 120 feet (36 metres) in height, are by no means the oldest trees in the garden. This honour belongs to the European larch, Norway spruce and mixed hardwoods planted by an earlier owner in 1820.

Although most of the estate is today managed by the Forestry Commission, 120 acres (48 hectares) to the west of the River Eachaig form the Younger Botanic Garden, an annexe of the Royal Botanic Gardens, Edinburgh.

The popularity of the garden rests on the success of the conifers, which grow rapidly in a place where the air sparkles and the annual rainfall can be as much as ninety inches (2,280 millimetres); and the rhododendrons, which provide colour for much of the year.

The vast number of rhododendrons at Benmore are directly attributable to the energies of three men: Sir Isaac Bayley Balfour, Regius Keeper of the Royal Botanic Gardens in Edinburgh at the turn of the century; George Forrest, plant hunter; and William Wright Smith, Balfour's chosen successor as Regius Keeper for thirty-four years from 1922.

Forrest, born in 1873 in Falkirk, was a natural explorer and adventurer. He began his working life in a chemist shop but his restless energy soon took him off to Australia. Although life in the outback suited his temperament, it failed to make his fortune and, on returning to Edinburgh, he took a job as a junior in the herbarium of the Royal Botanic Gardens. There he quickly absorbed a wide knowledge of the world's plants.

The move paid dividends when, in 1904, the Liverpool cotton-broker, Arthur Kilpin Bulley approached his friend Balfour for advice on who to send on a plant hunting

Above *Drama is the keynote of this garden, which is set on craggy slopes just south of Loch Eck.*

Left *Younger Botanic Garden, essentially an arboretum.*

expedition to China. For Balfour had no hesitation in recommending Forrest. From then until his death in Western China in 1934, Forrest sent back to the Edinburgh herbarium over 30,000 specimens and many hundreds of seeds.

Balfour had always been fascinated by rhododendrons and primulas and he had long nursed a dream of establishing in the west of Scotland a garden free from city pollution where he could grow the Sino-Himalayan specimens he so loved in conditions approximating their natural habitat. Sadly the dream never became a reality for him. He died in 1922 but his dream was instilled in his colleague and successor, William Wright Smith.

Smith needed little encouragement to study the plants of that particular region. He had earlier in his career spent several years at the Calcutta Botanic Garden and had been on many botanical explorations while living in India. When Younger presented his estate to the nation in 1928, Smith grasped the opportunity to create an annexe to the Royal Botanic Gardens.

Over the years the rhododendrons have flourished and some, such as the common *R. ponticum* planted by Duncan, have even become something of a pest, needing drastic attention. Now more choice species grow, such as *R. thomsonii* and *R. campanulatum*, self-seeding as are many of the other 250 species found in the garden. Needless to say, the display of colour between April and June is breath-taking.

Essentially the garden is an arboretum and its siting on craggy slopes just south of Loch Eck makes it the most exciting one in Great Britain. Overhead the buzzards soar lazily on the rising air currents, accentuating one's sense of isolation. It is an exhilarating place, where a journey can be taken from order to wilderness, as one moves outward from the central ordered core of the garden.

On the valley floor across from the fast-flowing River Eachaig, the avenue of massive Californian big trees, *Sequoiadendron giganteum*, make a bold statement, with their robust size and the way their trunks flare from the ground. Arranged in such rigid and yet harmonious order, the trees have an audacity unparalleled in the United Kingdom, and to stand alongside them is a humbling experience. The avenue runs across the valley floor to the very edge of the wild slopes. Nearby is the formal garden, a pocket of constrained nature ringed by a curtain of encroaching chaos, set against the slopes which contain a tangled mass of conifer, hardwood and rhododendron. There is a pond and an island planted with azaleas and hostas, the latter seeming somehow out of place. A stone cherub is locked in eternal combat with a dolphin fountain; both have a rather pathetic air, dominated as they are by the drama of the land and sky which surround them.

In spring primulas, Balfour's favourite, grow there and in autumn the sweet fragrance of *Eucryphia* x *nymansensis* washes the crisp air.

Drama is the keynote of this garden. Raw nature assaults one's senses at every turn. In no other garden in Great Britain is the air infused with such an intoxicating scent of pine, in no other garden does one appreciate its freshness so keenly. In such an environment the formal garden inevitably seems excessively manicured, its low terraced lawns edged by slow-growing conifers too precisely clipped and contrived. It is so clearly at odds with its surroundings, with the tree-clad slopes rising beyond the octagonal summer-house and with the slope on the other side of the valley, veined with the delicate tracery of waterfalls.

A path from the formal garden heads south-west, toying with the lower edges of the

The central ordered core of Younger Botanic Garden.

valley side until reaching the Golden Gates. These gates, brought by Duncan from the 1871 Paris exhibition, now mark the western extremity of the garden. The path skirts planting of many fine shrubs – magnolia, skimmia, rhododendron – before dog-legging behind Younger's former home, Benmore House, now an outdoor pursuit centre. Tall Douglas fir, *Pseudotsuga menziesii*, stand there and close by can be found the tallest western hemlock, *Tsuga heterophylla*, in Great Britain as well as the oldest trees in the garden, Scots pine planted in about 1820.

Along the path are dense thickets of rhododendron, including *R. hodgsonii*, a native of China, with pink and magenta blooms in April but attractive at almost any time of year with its pink-flushed bark. Lost among the shadows are *R. arboreum* and *Clethra barbinervis*, the white-flowering evergreen from China with bark resembling the eucalyptus but more subtly coloured. On all sides is a rich regenerative and calming silence, broken only by one's footfalls and the buzzards' shrill cries. Wellingtonias stand here, their tawny, stringy bark stained green by moss and highlighted by patches of silvery lichen.

From the Golden Gates a path leads up the valley side. Zig-zagging and excitingly slippery when wet, the path quickly attains a good height and, taken at a brisk stride, soon sends the blood racing through one's body. Staggering views are there for the taking, but there is beauty too, close at hand, in the mass of rhododendrons. *R. sinogrande* grows particularly well there, with its leaves up to two feet (sixty centimetres) in length.

The ultimate goal is the William Wright Smith memorial hut, perched high above the garden. It is a place in which to rest, drink in the wild beauty of this garden and contemplate the breath-taking views down the broad sweep of valley to Holy Loch and the distant Ayrshire hills. Here, truly, is paradise on earth.

Bibliography

AMHERST, ALICIA. *History of Gardening in England* (1895)

BALLARD, PHILADA. *An Oasis of Delight, The History of the Birmingham Botanic Garden*, Duckworth (1983)

CHAMBERS, WILLIAM. *Gardens and Buildings at Kew* (1763)

DREWITT, F. DAWTREY. *The Romance of the Apothecaries' Garden at Chelsea*, Cambridge University Press (1928)

EVELYN, JOHN. *Sylva* (1664)

FLEMING, LAWRENCE AND ALAN GORE. *The English Garden*, Michael Joseph (1979)

FLETCHER, HAROLD. *The Royal Botanic Garden, Edinburgh 1680–1970*, HMSO (1970)

GERARD, JOHN. *The Herbal* (1636 edition)

GUNTHER, R. *Oxford Gardens*, Oxford, Parker and Son (1912)

HADFIELD, MILES. *A History of British Gardening*, Hutchinson (1960)

HADFIELD, MILES. *Pioneers in Gardening.*

HEPPER, F. NIGEL (ed.). *Kew Gardens for Science and Pleasure*, HMSO (1982)

KADEN, VERA. *The Illustration of Plants and Gardens 1500–1850*, Victoria and Albert Museum.

LOUDON, JOHN C. *Encyclopaedia of Gardening* (1850)

PARKINSON, JOHN. *Theatrum Botanicum* (1640)

PREST, JOHN. *The Garden of Eden*, Yale University Press (1981)

WALTERS, STUART. *The Shaping of Botany at Cambridge*, Cambridge University Press (1981)

List of gardens generally not open to the public

Aberystwyth, University College of Wales Botany Garden, Ceredigion, Dyfed, Wales.

Birmingham, The University Botanic Garden, Winterbourne, Edgbaston, Birmingham, W. Midlands.

Coleraine, New University of Ulster, Londonderry, N. Ireland.

Enfield, University of London, Enfield, Middlesex.

Englefield Green, University of London Botanical Supply Unit, Englefield Green, Surrey.

Leeds, Botanical Experimental Garden, The University of Leeds, Leeds, W. Yorkshire.

Loughborough, The Arboretum, University of Loughborough, Charnwood, Leicestershire.

Reading, Plant Science Lab, The University of Reading, Reading, Berkshire.

Sheffield, Department of Botany Experimental Garden, University of Sheffield, S. Yorkshire.

Acknowledgements

The Author and Publishers would like to thank the following sources for allowing their photographs to be reproduced in this book:

A–Z Botanical Collection Ltd, 19, 140; Bath City Council 16 top and bottom; Biofotos – Heather Angel 67, 94, 148; Cambridge University Botanic Garden 38 left and right; Chelsea Physic Garden 44 top and bottom; Collins Publishers 9, 10, 24 top, 96; University of Durham Botanic Garden 59 top and bottom, 60, 61; Royal Botanic Garden, Edinburgh 152, 155; Fletcher Moss Botanical Gardens 63; Ben Gibson 145; City of Glasgow Parks and Recreation 69; Harlow Car Gardens 72; Keith Anderson, University of Hull 77; Impact Photos – Pamla Toler 89, 91, 108; Royal Botanic Gardens, Kew 114, 119; The University of Liverpool Botanic Gardens 90; Tania Midgeley 48, 49; J. K. Burras, Oxford University Botanic Garden 97, 100; Photos Horticultural – Michael and Lois Warren 1, 21, 28, 30, 32, 34, 35, 37, 43, 50, 53, 57, 70, 74, 92, 104, 116, 117, 129, 132; Kenneth Scowen 2–3, 138, 144; Sheffield Botanical Gardens 123, 125; The South London Botanical Institute 81; University College of Swansea Botanic Garden 136; Michael Young 23, 24 bottom, 27, 54, 84, 153.

Index

Abies grandis 141
  koreana 57, 64
  magnifica 22
Acacia crassiuscula 110
  dealbata 51, 130
  spadicigera 111
Acaena inermis 73
Acer capillipes 150
  griseum 29, 91, 129
  grosseri 29
  japonicum 'Aconitifolium' 85
    'Aureum' 85
  palmatum 17, 85, 150
  palmatum 'Atropurpureum' 62, 130
    'Dissectum Ornatum' 18
    'Senkaki' 33, 91
  pensylvanicum 57
  platanoides 85
    'Schwedleri' 75
Achillea ptarmica 18
Aconitum japonicum 53
Acorus calamus 139
  'Variegatus' 40
Actinidia chinensis 129
  kolomikta 63, 100
Aesculus × carnea 86
Agave recurvata 82
Ailanthus altissima 33, 42, 50
Aiton, William 114
Aiton, William Townsend 115
Akebia trifoliata 100
Alnus cordata 75
alpines 26, 41, 62, 87, 120
Alsophila australis 77
Alyssum saxatile 63, 64
Amsonia tabernaemontana 129
Androsace 64
Anemone nemorosa 108
Anthemis tinctoria 18
Aponogeton distachyus 27
Apothecaries, Society of 45, 46
Araucaria araucana 58
arboretum 52, 68, 71, 75, 105, 108, 112
Arbutus meziesii 100
Arenaria balearica 73
Aristolochia brasiliensis 77, 110
  elegans 130
Aristotle 7
Artemisia schmidtiana 'Nana' 107
Arum italicum 33
  maculatum 30, 99
Arundinaria murieliae 139
  nitida 99
Arundo donax 29, 101
Asarum europaeum 82, 128
Asperula lilaciflora 87
Asphodeline lutea 128
Asplenium scolopendrium 73
Astrantia major 18
Atropa belladonna 34, 109
Augusta, Dowager Princess of Wales 113, 114, 115
Ayliffe, Dr John 98

Balfour, Dr Andrew 9
Balfour, Isaac Bayley 103, 105, 106, 107, 151, 153
Balfour, John Hutton 105
Banks, Sir Joseph 11, 47, 115
Baskerville, Thomas 95, 97, 98
Bath Botanical Gardens 16, 17–18, 19
bear pit 125
Bedgebury National Pinetum 20–22, 21, 23
Bees Ltd 88, 106
Begonia foliosa 69
Bell, Prof. Arthur 13, 14
Berberis wilsoniae 141
Bergenia ciliata 50
Betula papyrifera 29
Bicton 11
Birmingham Botanical and Horticultural Society 25
Birmingham Botanical Gardens 11, 12, 24, 25–30, 27, 28, 30
Blechnum spicant 73
Bligh, Captain 115
Bobart, Jacob 95, 96
Bobart, Jacob, the second 97, 98
boscage 118
Botanic Magazine 48
bougainvillea 41
Bougainvillea glabra 57
Boucher and Cousland 68
Brassia 109

Brightman, Frank 80, 82
Bristol University Botanic Gardens 31–5, 32, 34, 35
Broome, C. E. 17
Brown, Capability 115, 121
Brunnera macrophylla 76
Bulley, Arthur Kilpin 88, 106, 151
Burras, Ken 14
Burton, Decimus 11, 115, 116, 119
Bute, Lord 113, 114
Butomus umbellatus 139

Calcutta Botanic Garden 26
Calliandra haematocaphala 30
Callistemon 110
Calocedrus decurrens 128
Caltha palustris 31, 64
Cambridge 46
Cambridge Flora 36
Cambridge University 8
Cambridge University Botanic Garden 10, 12, 36–43, 37, 38, 43
Capel, Sir Henry 113
Cardamine pratensis 141
Cardiocrinum giganteum 133
Carica papaya 61
Carpinus betulus 129
Cassinia fulvida 75, 85
Catalpa bignoniodes 'Aurea' 18, 78
  × *erubescens* 'Purpurea' 41
Caucasian speedwell 33
Ceanothus dentatus 100
  'Gloire de Versailles' 63
Cedrus atlantica 33
  'Glauca' 40, 85, 86, 150
  deodara 26, 33, 86
  libani 46, 86
Cephalotaxus 22
Ceratostigma plumbaginoides 85
Cercidiphyllum japonicum 27
Cereus peruvianus 57
Chad Valley 25
Chaenomeles 51
Chamaecyparis lawsoniana Hazelmere Form 22
  thyoides 22
Chambers, Sir William 113, 114
Charles II 96
Chatsworth 11, 26
Chelsea 9, 10, 11, 12, 13, 26
Chelsea College 13
Chelsea Physic Garden 44, 45–51, 48, 49, 50
Cheyne, Charles 46
China 18, 139, 153
chronological borders 42, 43, 68
Chrysanthemum maximum 100
Chusan palm 136, 139
Cibotium glaucum 111
Cinchona officinalis 117
Cinnamomum camphora 109
Cistus 51
Clematis 'Lincoln Star' 86
  Marcel Moser 86
  tanguitca 86
Clerodendron trichotomum 30
Clethra barbinervis 155
Colchicum 34, 53
conservatory 65, 81, 122
Cook, Captain 47, 115
Cooper, Roland 88
Cordyline australis 93
Cornus alba 107
  kousa 60
  mas 'Variegata' 17
  nuttallii 63
  stolonifera 'Flaviramea' 123
Corokia cotoneaster 87
Corylopsis spicata 91
Cotinus coggygria 61, 126
Cotoneaster horizontalis 107
  lactea 136
Crinodendron hookerianum 93, 130
Crinum moorei 41
Cruickshank, Anne 52
Cruickshank Botanic Garden 52–4, 53, 54
Culpeper, Sir Edward 137
Cupressus cashmeriana 78, 110
Curtis, Eric 68, 69
Curtis, William 48
Cyathea australis 120
  medullaris 87
Cyclamen hederifolium 91
Cyperus papyrus 29
Cyphomandra befacea 110
Cytisus ardoinii 107
  battandieri 30

Dactylorrhiza elata 73, 108
  fuchsii 82
Danvers, Lord (Danby, Earl of) 8, 95
Daphne tanguitca 50
Darwin, Charles 38
Datura rosea 30
  sanguinea 30
  stramonium 109
Daubeny, Dr Charles 98, 99
David, Armand 105
Davidia involucrata 101, 105, 139, 150
Dawyck Botanic Gardens 112
De Materia Medica 7
Dendrobium 109
Diascia rigescens 94
Dichorisandra thyrsiflora 29
Dicksonia arborescens 29
  antarctica 29, 68, 77, 93, 111, 120
  × *Lathamii* 29
Digitalis 34, 49
Dill, Dr Johann (Dillenius) 98
Dioscora bulbifera 110
Dioscorides 7
Dipelta floribunda 18, 100
Dodecatheon meadia 64
Dodoens, Rembert 8
Douglas, David 143
Dracaena draco 130
Drosera rotundifolia 42
Dryas octopetala 133
Dryopteris affinis 101
  felix mas 55
Duncan, Donald 13
Duncan, James 151, 154
Dundee University Botanic Garden 55–7, 56
Durham University Botanical Garden 58–61, 59, 60, 61

Eachaig River 151, 154
Echium pininiana 94
Eck Loch 156
ecological areas 55
Edinburgh 9, 26, 57, 107
Ellis, Humphrey 95
Encephalartos longifolius 119
Endeavour 47
Eremurus robustus 52
Eucalyptus niphophila 130
  perriniana 130
Eucryphia × *nymansensis* 149, 154
Euonymous alatus 85
Euphorbia lathyrus 29
  robbiae 50
  sikkimensis 123
Evelyn, John 9, 20, 46, 96, 113, 125, 143

Fagus sylvatica purpurea 87
Fagus sylvatica 'Riversii' 149
Festuca glauca 73
Fletcher Moss Botanical Gardens 62–4, 63
Fontrey, Samuel 118
Forestry Commission 20
Forrest, George 88, 105, 106, 112, 151, 153
Forsyth, William 47
Fortune, Robert 48
Fox, Dr Wilfred 147, 149, 150
Frederick, Prince of Wales 113
Fremontia californica 51

Galtonia candicans 53
Gardener's Magazine 11
Gardenesque 83, 85, 121, 122
Garrya elliptica 93
Genista aetnensis 77
Gentiana septemfida 91
George II 113
George III 10, 47, 113, 115, 120
Georgia 47
Geranium dalmaticum 43
  himalayense 'Gravetye' 94
  macrorrhizum 52, 99
  phaeum 50
  pratense 50
  procurrens 99, 139
  psilostemon 99
  sanguinneum 76, 128
  versicolor 99, 101
Gerard, John 8, 9, 38, 42, 45, 51, 53, 55, 81, 82, 101, 124, 149

Gilmour, John 42
Gladiolus papilio 73
Glasgow Botanic Gardens 11, 25, 65–9, 67, 69
glasshouses 29, 34, 35, 41, 42, 46, 51, 52, 54, 57, 61, 76, 77, 87, 119, 120
Glastonbury Leechdome 7
Glaucium flavum 41
Graham, Dr Robert 65
Great Fire (1666) 46
Griselinia littoralis 94, 136
Gunnera manicata 27, 107
Gymnocladus dioica 99

Harcourt, Edward 102
Harlow Car Gardens 70, 71–5, 72, 74
Harrison, Stockdale 83
Harrison, William 8
heather gardens 87, 90
heathers 43, 52, 55, 60, 64, 73, 74
Hebe 'Carl Teschner' 85
Hedychium 110
  gardnerianum 30
Helleborus corsicus 29, 50, 51, 53, 85
  niger 133
Henderson, Prof. Douglas 93
Henry, Augustine 105, 106, 139
Henslow, John Stevens 11, 12, 13, 38, 39, 40
herbarium 80, 81
herbaceous borders 17, 18, 29, 33, 41, 42, 52, 58, 85, 90, 101, 109, 124, 131
herb border 76, 78
herb garden 68, 83, 86
herbs 57, 75, 109
Herman, Paul 46
Hesperis matronalis 85
Hevea brasiliensis 116
Hibiscus rosa-sinensis 41, 77
Hieracium maculatum 54
Hobson's Conduit 40
Holford, Sir George 143
Holford, Robert Stayner 143
Holy Loch 156
Holyrood Abbey 10
Hooker, Sir Joseph 115, 116
Hooker, William Jackson 66
Hope, John 103
Hopkirk, Thomas 65
Horticultural Society 47
Hosta sieboldiana 129
Hull University Botanic Garden 25, 76–8, 77
Hume, Allen Octavian 79
Hume's South London Botanical Institute Garden 79–82, 81
Hydrangea aspera macrophylla 77
  paniculata 149
  quercifolia 40
  sargentiana 77
  serrata 'Thunbergii' 77
Hyoscyamus niger 109

Indian National Congress 79
Inverleith 105
Ipomoea acuminata 29
Iris pallida 53
  reticulata 91
  sibirica 33, 73

Jekyll, Gertrude 40
Johnson, Thomas 9, 45, 46, 51
Jones, Inigo 95
Jubaea chilensis 120
Juglans nigra 86
Juniperus 22, 55, 107
  procumbens 133
  squamata 'Meyeri' 75

Kalm, Peter 47
Kent, William 126
Kibble, John 66, 67
Kibble Palace 11, 65, 68
Killarney fern 48
King's College Chapel, Cambridge 36
Knaresborough Forest 71
knots 118
Kolkwitzia amabilis 60

Latham's tree fern 29
Latham, William Bradbury 29
Lathraea clandestina 40, 73
Laurus nobilis 94
Leicester University Botanic Garden 83–7, 84
Leiden, University of 46
Leith Walk 105
Lewisia columbiana 'Wallowensis' 62
  tweedyi 62
Leycesteria formosa 27
Leyden 47
Libocedrus decurrens 64
Ligustrum lucidum 61
Lilium mackliniae 63
Lillie, Beatrice 147
limestone rock garden 41
Limonium latifolium 76
Linnaeus 47
Liquidambar styraciflua 18, 20, 34
Liriodendron tulipifera 33, 64, 135
Liverpool University Botanic Garden 11, 12, 25, 88–91, 89, 90, 91
Livistona australis 109
Lobelia cardinalis 40
  speciosa 40
Loder, Gerald 137
Logan Botanic Garden 92, 93–4, 94
London, George 38
Loudon, John Claudius 11, 26, 85, 112, 121, 143
Lynch, Richard Irvine 39
Lysichiton americanus 64, 73
Lysimachia 33
Lythrum salicaria 18, 33, 53, 58

Magnolia acuminata 26
  campbellii 139
  dawsoniana 18
  grandiflora 85
  soulangeana 50, 99
  stellata 18, 50
  tripetala 18
  wilsonii 137
Maianthemum kamtschaticum 108
Manchester 25, 62
Marnock, Robert 121, 122
Matteuccia struthiopteris 55
McDouall, Kenneth and Douglas 94
McNabe, William 105
Meconopsis betonicifolia 34
  cambrica 40, 62, 73
  grandis 94
Mentha requienii 133
Metasequoia glyptostroboides 18, 22, 58, 62, 73, 128
Mickwell Brow 88
Miller, Philip 10, 47, 51
Mimosa pudica 41
Mirabilis jalapa 82
Miscanthus sinensis 'Variegatus' 128
Mitchell, Bob 13, 14, 131
Monstera deliciosa 30
Montpellier 8
Morrison, Robert 96
Morus nigra 18
Murray, Patrick 103
Murray, Stewart 65
Myrrhis odorata 100

Naesmyth family 112
Natural History Museum, London 79
Nelson, David 115
Nepenthes × Ratcliffeana 110
Nesfield, William Andrew 115
Nicandra violaceae 109
Nicotiana affinis 135
Nigella damascena 53
  hispanica 41
Northern Horticultural Society 71
Nothofagus obliqua 74
Nuneham Courtney 102

Omphalodes cappadocica 54
Onoclea sensibilis 133
Onopordum acanthium 76

Ophioglossum vulgatum 87, 108
Osmunda regalis 64, 139
Oxford 8, 9, 13, 14
Oxford University Botanic Garden 95–102, 96, 97, 100

Pachysandra terminalis 139
Padua 89
Papaver nudicaule 129
Paris 8, 47
Parkinson, John 10, 42, 43, 54, 101, 133
Parrotia persica 146
parterres 51, 86, 118, 126
Passiflora antioquiensis 35
Paulownia lilacina 136
  tomentosa 42, 49
Paxton, Joseph 11, 26, 122
peat gardens 107, 108
  terraces 73
pergolas 86, 135
Petasites albus 101
  japonicus 86
Phallus impudicus 60
Pharbitis learii 110
Philadelphus coronarius 61
Phillyrea angustifolia 64
Phormium tenax 64
  'Purpureum' 64
Phlox douglasii 63
  subulata 63, 64, 91
Physicians, College of 46
Picea 22
  abies 141
  Nudiformis 73
  glauca Albertiana cornica 64
  pungens glauca 62
  pungens Koster 62
  pungens glauca pendula 107
  orientalis 22
picturesque 121
Pieris formosa forrestii 91, 129
Pinus coulteri 126
  nigra 29, 101
  pinea 130
  sylvestris 20, 73, 87, 90
Pisa Medical Garden 125
Pittosporum tenuifolium 94
Platycerium bifurcatum 87
Pleione bulbocodioides 91
Pliny 7
Plumbago capensis 35
Plumeria rubra 110
Polygonum 81
  affine 33, 133, 141
  amplexicaule 53
  bistorta 81
  cuspidatum 101
  hydropiper 81
  persicaria 81
  sachalinense 101
  vacciniifolium 139
  vulgare 55
Polypodium subauriculatum 78
Polystichum setiferum divisilobum 73
Populus lasiocarpa 14
Price, Sir Henry 137
Primula flaccida 91
  florindae 18, 107
  helodoxa 64
  ioessa 63
  scotica 91
Prunus serrula 91, 123
  shirotae 17
  × *yedoensis* 50
Pseudotsuga menziesii 145, 155
Pterocarya fraxinifolia 40
Pulsatilla 129
  pratensis 62
  vulgaris 41
Puya alpestris 69

Queen's Garden, Kew 118
Quercus ilex 49, 90
  robur 22
  rubra 22
  suber 51

Ravenala madagascariensis 30
Ray, John 10, 36, 38, 46
Rendle, A. B. 82
Repton, Humphrey 121, 122
Reynoutria japonica 64
Rheum palmatum 128
Rhinanthus minor 87
Rhododendron arboreum 155
  augustinii 52
  auriculatum 22
  barbatum 137, 139
  campanulatum 154
  campylogynum 52
  delavayi 139
  falconeri 139
  forrestii 107
  hodgsonii 155
  macabeanum 63
  mucronulatum acuminatum 90
  ponticum 154
  rex 75
  × *shilsonii* 137
  sinogrande 139, 156
  succothii 139
  thomsonii 154
rock garden 26, 31, 39, 41, 47, 52, 54, 62, 64, 73, 75, 90, 101, 107, 118, 133, 137
Rodgersia podophylla 18
Romneya coulteri 30
Rosa hugonis 63
rose garden 124
Royal Botanical Institution 65
Royal Botanic Garden, Edinburgh 103–11, 104, 108
Royal Botanic Gardens, Kew 11, 12, 13, 14, 20, 26, 47, 66, 68, 113–20, 114, 116, 117, 119
Royal Society 46
Royal Victoria Park, Bath 17
Rubus biflorus 43
  cockburnianus 50
  thibetanus 81
Ruta graveolens 41

Sabal blackburneana 109
Saccharum officinarum 29
Saint Helena Ebony 69
Salix gracilistyla 33
  rosacea 82
Salvia horminum 53
  superba 33
Sarcococca confusa 50
  hookeriana 40
Sarracenia 42
Saxegothaea conspicua 22
Saxifraga 64
  cespitosa 91
Senecio squalidus 98, 100
Sequoia sempervirens 141
Sequoiadendron giganteum 18, 22, 33, 145, 151, 154
Sheffield Botanical Garden 121–7, 123, 125
Sherard, Dr William 98
Sibbald, Robert 9, 103
Sieveking, Albert Forbes 12
Silene acaulis 133
Silybum marianum 52
Simmondsia chinensis 82, 110
Skimmia, National Collection of 86
Smilacina racemosa 49, 128
Smith, William Wright 106, 107, 151, 154, 156
Solanum crispum 100
Sorbus Mitchelli 85
  prattii 129
  'Wilfred Fox' 149
Southampton University Botanic Garden 128–30, 129
Stewartia pseudocamellia 149
St Andrews University Botanic Garden 13, 131–4, 132
Stone, Neklaus 95
Strelitzia nicolai 120
  reginae 35, 41, 87, 101
Sutherland, James 103
Swansea, University College of, Botanic Garden 135–6, 136

Swan Walk 47
Sylva 20, 96
Syringa × persica 'Alba' 100

Tanacetum densum 41
Taxodium distichum 22, 85
Taxus baccata 49
Tay, Bridge 55
Tay, Firth of 55
Taylor, Benjamin Broomhead 122
Taylor, Sir George 118
Tellima grandiflora 99
Temple, Sir William 12, 113
Tetracentron sinense 40
Thrysopteris elegans 68
Thuja 22
Thunbergia grandiflora 110
Thuya occidentalis 76
Tibouchina urvilleana 29, 57
Tilia × europaea 40
topiary 118
Trachycarpus fortunei 41, 63, 77, 93, 136
Tradescantia 33
Tradescant, John 95
Trichomanes speciosum 48
Trillium grandiflorum 34, 49, 108, 130
Tropaeolum polyphyllum 133
Tsuga heterophylla 139, 155
Typha angustifolia 139

Uffenbach, Conrad von 48, 97
Ulex europaeus 55
Utrecht University of 82

Vaccinium myrtillus 55
Veitch, Sir John Henry 105, 139
Verbascum thapsus 61, 127
Verey, Rosemary 82
Viburnum 'Anne Russell' 51
  × *bodnantense* 43
  carlesii 64
  tinus 49
Victoria amazonica 26
  cruziana 57, 69
Victoria, Queen 116
Violas, National Collection of 86
Vitis coignetiae 76, 135
  vinifera 86
Vriesea imperialis 110

Wakehurst Place 137–42, 138, 140, 142
Walker, Richard 10, 38
Walpole, Horace 113
Ward, Frank Kingdom 63, 88
Wardian Case 48, 50
Ward, Nathanial 48
Watts, John 46, 47
Weigela florida variegata 61
Westonbirt Arboretum 143–6, 144, 145
wild flower garden 62
Williams, Don 155
Williams, John Charles 106
Wills, Capt. Douglas 31
Wills, Melville 31
Wilson, Ernest Henry 105, 106, 112, 139
Wilson, Dr John 131
Winkworth Arboretum 147–50, 148
winter garden 43, 149
Winterton, William 83
Wisteria sinensis 86
Wright, Henry 71

Xanthoceras sorbifolium 100
Xanthosoma violaceum 57, 110

Younger Botanic Garden 106, 151–5, 152, 153, 155
Younger, Harry George 151
Yucca gloriosa 18, 40, 43